DATE			

WESTERN SUNRISE

Titles of related interest

Collapse and survival
R. Ballance & S. Sinclair

Electronics and industrial policy
S. Jacobsson

High tech America
A. Markusen, P. Hall & A. Glasmeier

New firms: an economic perspective
P. Johnson

Planning in Europe
R. Williams (ed.)

Production, work, territory
A. Scott & M. Storper (eds)

Regional dynamics
G. Clark, M. Gertler & J. Whiteman

Regional economic development
B. Higgins & D. Savoie (eds)

Silicon landscapes
P. Hall & A. Markusen (eds)

Technological change, industrial restructuring and regional development
A. Amin & J. B. Goddard (eds)

Urban and regional planning
P. Hall

WESTERN SUNRISE

The genesis and growth of Britain's major high tech corridor

—————

Peter Hall · Michael Breheny
Ronald McQuaid · Douglas Hart

London
ALLEN & UNWIN
Boston Sydney Wellington

Allen & Unwin, the academic imprint of

Unwin Hyman Ltd

PO Box 18, Park Lane, Hemel Hempstead, Herts HP2 4TE, UK
40 Museum Street, London WC1A 1LU, UK
37/39 Queen Elizabeth Street, London SE1 2QB

Allen & Unwin Inc.,
8 Winchester Place, Winchester, Mass. 01890, USA

Allen & Unwin (Australia) Ltd,
8 Napier Street, North Sydney, NSW 2060, Australia

Allen & Unwin (New Zealand) Ltd in association with the Port
Nicholson Press Ltd,
60 Cambridge Terrace, Wellington, New Zealand

First published in 1987

British Library Cataloguing in Publication Data

Western sunrise: the genesis and growth of
Britain's major high tech corridor.
1. High technology industries – Great Britain
I. Hall, Peter, 1932–
338.4'76'0941   HC260.H53
ISBN 0–04–338142–1

Library of Congress Cataloging-in-Publication Data

Western sunrise.
Bibliography: p.
Includes index.
1. High technology industries – Government policy –
Great Britain.   2. High technology industries – Great
Britain – Location.   I. Hall, Peter Geoffrey, 1931–
II. Title.
HC260.H53W47   1987   338.6'042   86–28755

Set in 10 on 11 point Palatino by Columns of Reading
and printed in Great Britain by
Mackays of Chatham

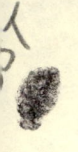

Dedication

David Bishop, co-director of the Joint Centre for Land Development Studies of the University of Reading and the College of Estate Management, died after a long illness just as this book was going to press. He was 42.

He played a major role in initiating the research on which this book was based. The extent of his contributions – both intellectual and administrative – to this and to countless other research projects is known only to those who worked with him day by day.

He was never too busy to help others in their work; to lighten their administrative loads; to deal constructively and unobtrusively with their research problems. We and our co-researchers at Reading owe him more than we can say.

Applied urban research has lost a colleague and friend whose greatest contributions were yet to come. We dedicate this book to his memory.

Preface and acknowledgements

This book is based on research supported by a grant from the Environment and Planning Committee of the Economic and Social Research Council to the University of Reading. The research project, which ran from October 1983 to February 1985, was initiated by the Joint Centre for Land Development Studies and conducted in the Department of Geography.

On the university campus, four collaborators deserve particular mention: David Bishop, co-director of the Joint Centre, who helped initially to administer the project and to whose memory this book is dedicated; Paul Cheshire, originally a co-principal investigator, who powerfully helped shape the research design; Robert Langridge, an ESRC linked award graduate student, who helped to produce the definition of high technology industries reported in Chapter 2; and Paschal Preston, who was research officer for the first six months.

Off-campus, the contributors are too numerous to thank individually. But certain individuals and groups require special mention: Berkshire County Council's Planning Department, who made available unpublished survey material; the many owners and managers of Berkshire high tech firms who gave valuable time for extensive interviews; the Manpower Services Commission for making available their NOMIS data base. Without these helpers the book could not have been written; we gratefully acknowledge them. For errors of fact or interpretation, however, the authors are solely responsible.

In the writing of a book with four authors, it is inevitable that some division (not necessarily spatial) of labour is involved. The particular labours ought to be acknowledged.

The original research proposal was drafted by Peter Hall, Michael Breheny, Douglas Hart and our colleague Paul Cheshire. Our research student Robert Langridge also helped in discussions at this stage. These early views are contained in Breheny *et al*. (1983). Paschal Preston was research officer on the project for the first six months. Ronald McQuaid took over at this point and saw the funded part of the project through to its conclusion in the spring of 1985. Much of the basic empirical work was carried out by him. Michael Breheny was responsible for the day-to-day management of the project.

In the writing of the final text initial drafts were prepared by different authors. Peter Hall took responsibility for the work on high tech company interviews, theoretical perspectives, the origins of the electronics industry, the role of government research establishments, and public infrastructure. Michael Breheny was responsible for the two chapters on the geography of high technology industry, Ronald McQuaid, Peter Hall and Michael Breheny worked on the defence chapter, Ronald McQuaid and Michael Breheny on the definitions chapter. Peter Hall and Douglas Hart researched and wrote the chapter on the role of planning in the development of high tech industry to the west of London. Peter Hall produced initial drafts of the introductory and concluding chapters; Douglas Hart and Michael Breheny produced the final drafts. Michael Breheny and Douglas Hart prepared the final draft of the book for publication.

Peter Hall, Michael Breheny, Ronald McQuaid and Douglas Hart

Contents

PART I   *What?*

PART II   *Where?*

PART III *Why?*

PART IV *So what?*

List of Tables

List of Figures

PART I *What?*

1

Corridors of power

In the 1980s, throughout the advanced industrial nations, policy makers and opinion formers have seized upon a new recipe for economic regeneration and a new image of economic success. The recipe is high technology industry; the image, the corridors of growth where high tech industry clusters. High tech is presented as the answer to the decline of older smokestack or sunset industries, once themselves at the technological frontier, now too easily imitable by newly industrializing countries. High tech growth areas, in the media treatment, even rank their own neologisms: 'silicon landscapes'. California's Silicon Valley, first and still most famous, is joined by Silicon Prairie in Texas, Silicon Glen in Scotland, and Silicon Fen around Cambridge (Hall & Markusen 1985). And the M4 Corridor, supposedly running the length of the motorway across southern England from west London to Bristol and South Wales, is hailed by the media as Britain's home-grown version of Silicon Valley. It is 'Sunrise Strip', 'Britain's California within the home counties', the place where 'the recession passed almost unnoticed', 'the focus of the next industrial revolution' (*Economist* 1982, Walker 1982, Thomas 1983b, Feder 1983, Neil 1983, Phillips *et al.* 1983).

In all this, there has been much 'hype' and little hard assessment. It is not always even clear what exactly high tech is, beyond the fact that it is associated with computers. The number of jobs which high tech industry, however defined, has actually created remains obscure. It is not certain that phenomena like the M4 Corridor are much more than convenient vehicles for features in the business pages or for estate agency promotions. In other words, it awaits to be established what high tech is, where it is, and where it has been growing. And, since the basic facts are not known, it is idle speculation to ask about the reasons why.

In the United States, a little more has been achieved (Markusen *et al.* 1986). There have been studies of single industries, single areas or single aspects: of individual high tech industries (software, semi-conductors, robotics, biotechnology) in California; of the genesis and growth of Silicon Valley; of the geography of research and development (R&D). Such studies give valuable insights: about the often-astonishing early growth rates of new high tech industries (Hall *et al.* 1983, 1985); about the critical roles played in Silicon Valley by Stanford University, its professor of electrical engineering (Frederick Terman) and Pentagon contracts (Saxenian 1981, 1985a); about the different location patterns of industrial and government R&D, the first in the old-industrialized Frostbelt, the second in the Sunbelt (Malecki 1981b). And we now have a systematic study of the geography of high tech across the entire United States, and of the reasons for it, which questions some popular myths (Markusen *et al.* 1986).

No such general account exists for Britain; even partial studies are few. There are two valuable papers on the geography of British R&D (Buswell & Lewis 1970, Howells 1984), which emphasize its extraordinary concentration in the South-East. Related to this, there is a study of the geography of innovating firms, which – perhaps unsurprisingly – finds that these too are highly concentrated in the South-East and in the neighbouring South-West and East Anglia regions (Oakey *et al.* 1980). There is work on the electronics industry (Oakey 1981, 1983). But there is no general geography of high technology industry in Britain; nor is there any study of its concentration – whether real or mythical – in the M4 Corridor.

This book, and the study on which it was based, represents an attempt to remedy both these deficiencies. In our work, we first asked some fairly basic questions of the *what?* variety. What exactly do we mean by high tech industry? Is high tech growing, and if so by how much? Then we turned to *where?* questions. Where in Britain do high tech industries concentrate, if they concentrate at all? Does this concentration take the form of a corridor, or some other shape? Within this area of concentration, what kinds of firms produce high tech products, how do they start, and how do they come to be there?

Then we logically began to ask *why?* questions. What kind of theory would best explain such high tech industrial concentrations? What hypotheses might we develop from the theory, to test against the evidence? Was the reason simply out-migration of industry from historic high tech clusters within London? Or is there an additional factor – such as the presence of Government Research Establishments (GREs) engaged in defence research? What is the nature of the links between the GREs and the firms? Do public planning policies, or the provision of infrastructure, have any effect on the growth of high tech?

Posing questions in this order makes it seem very tidy – after the event. Real-life research is messier than that. We had more than the usual share of definitional headaches and of grappling with statistical data, not least in the elusive field of defence. But we received abundant help from many people, including the proprietors and managers of the many firms we interviewed.

A guide through the book

Chapter 2 begins answering the *what?* questions by a basic definitional deck-clearing operation. Following the parallel American work (Markusen *et al.* 1986), it looks at alternative definitions of high technology industry and concludes that one based on the scientific or technological character of the labour force is best. Accordingly, it produces two listings: a long one, corresponding as precisely as possible to the American equivalent, and a shorter one, encompassing the most important industries and thereby the great majority of the high tech workforce. For convenience, it is this latter definition that is mainly used in what follows.

Chapters 3 and 4 really constitute a single body of analysis, seeking to establish the facts about high tech jobs and where they are located. Chapter 3 continues to ask *what?* questions by first looking at the overall national pattern of job change in the 1970s and early 1980s. It comes to a surprising – and extremely depressing – conclusion about the role of high tech in job creation. From this, it goes on to *where?* questions, posed at the national level: where in Britain does high tech locate? The biggest *absolute* concentrations, in terms of sheer numbers of workers, unsurprisingly tend to be in the largest and most populous places; these are not necessarily the top concentrations in terms of the *relative* importance of high tech industry.

Combining these alternative measures, Chapter 3 concludes that there is one major high tech concentration, and two minor ones. The minor ones are in north-west England and in central Scotland ('Silicon Glen'). But they pale into insignificance in comparison with the major one, which comprises Greater London and the counties that wrap around it to the south-west, west and north: a Western Crescent, rather than a Western Corridor. Nevertheless, the biggest and fastest-growing concentration within this Crescent does lie to the west of London. Thus, we do have an M4 Corridor, but a short, concentrated one – focused on Berkshire and north Hampshire – superimposed on a Western Crescent of economic growth. The analysis presented here does *not* identify the long M4 Corridor – from west London to South Wales – that is popularly assumed.

Chapter 4 looks in greater detail at both Crescent and Corridor. It draws on a survey by the Royal County of Berkshire, in the centre of the M4 Corridor, to establish a profile of high tech firms and to show how activity in the area has grown since World War II. It shows that indeed the Corridor is the heart of the wider Crescent, in terms of both scale and function. It has a concentration of high tech jobs, and those jobs tend to be in administrative and research headquarters units; manufacturing is secondary.

Chapter 5 builds on this geographical analysis, describing the results of interviews with a large number of high tech firms in Berkshire, which formed a focal point of our study. It reports what these firms told us about themselves and in particular about the reasons that had brought them to this location. Most, it emerges, are in Berkshire because that is where they were started. And, though many now say that airport or motorway matters to them, it is clear that this cannot have been the initial factor in location. That, in many cases, was the GREs.

Chapter 6 marks a distinct break. Now that the basic facts about high tech location have been established, it begins to ask the *why?* questions. It looks at textbook theories of location – both those in the older neo-classical tradition, and those in the newer political economy tradition – to ask what light they throw upon the high tech phenomenon. It finds them helpful in some particulars, unsatisfactory (because incomplete) in others. Accordingly, the chapter concludes, we must construct an eclectic explanation that combines parts of several building blocks. Particularly, this explanation must account for the changing geography of innovation: why certain areas, at certain times, become seedbeds for clusters of firms, many of them new, using scientific knowledge to develop new industrial traditions.

Chapter 7 is the first of a set of chapters that try to use these insights to build such an eclectic explanation of innovation for the Crescent and Corridor. It examines an initial hypothesis that these areas represent nothing but the physical displacement of a much older seedbed, located in the heart of London in localities such as Clerkenwell and Pimlico: the homes of London's long tradition in the manufacture of what were once known as scientific and philosophical instruments. As these trades grew and evolved into the electrical and then electronic industries, so, in the search for space, they migrated outwards; in the 1920s and 1930s to the north-west sector of Greater London, in the 1950s and 1960s to the home counties beyond. This process, the chapter concludes, provides at least an important part of the explanation. But it fails to establish why the resulting geographical pattern should have such a pronounced western bias.

For this, Chapter 8 provides another explanation: the concentration west of London, already well-established before World War II, of the Government Research Establishments (GREs). These, overwhelmingly concerned with defence or with fundamental research with defence applications, were strongly localized in the sector between Portsmouth on the south coast and the old Great West Road. World War II, and the cold war period of the early 1950s, saw a great expansion in their activities – as well as a geographical spread, with the establishment of the complex of nuclear research in the Didcot–Harwell–Culham triangle south of Oxford.

Chapter 9 takes up this clue, seeking to trace how, during the 1950s and 1960s, the GREs developed a complex web of industrial procurement. Overwhelmingly the initial beneficiaries were large and well-established firms in the new electronics industry, which had developed out of the radio industry of the 1930s and had expanded massively during World War II. In turn they subcontracted specialized work to smaller companies. The nature of this contracting and subcontracting process was such as to lock the firms into close proximity with the GREs themselves, producing a curious high tech analogue of the old inner-city industrial quarter, wherein each firm depended on orders from the wholesale house. But now, since the GREs had rural locations, the industry necessarily followed them.

The two succeeding chapters seek to follow this lead. If public policy, through the GREs, provided one crucial basis for the development of a high tech industrial base, did it also contribute in other ways? Chapter 10 looks at the role of land-use planning from the 1944 Greater London Plan onwards. It concludes that in the Crescent and Corridor, the generally negative and restrictive nature of the planning policies paradoxically enhanced the area's high tech role, by maintaining the high quality of the natural and built environment.

Chapter 11 similarly looks at the role of infrastructure, especially the area's national and international transportation links. It concludes that the three most important – London's Heathrow Airport, the M4 motorway itself and the high-speed rail system – were all planned and developed for quite different reasons, quite unconnected with any notion of facilitating high tech industry. That they may have played a subsequent role is fortuitous, especially since most firms established themselves in the Corridor before either motorway or high-speed rail even existed.

Chapter 12 sums up. It starts by summarizing the main conclusions of the book: its answers to the *what?*, *where?* and *why?* questions. Then it goes on more speculatively, to discuss some of the wider implications: the *so what?* questions, which cannot be answered by research but may be at least addressed through speculation. First, it tries to set high technology within a wider framework of the sources of employment growth. To be sure, high tech has contributed to the present prosperity of Crescent and Corridor – but it has been by no means the sole factor, or even the most important. To a greater extent than many other regions of Britain, the home counties west of London now have a post-industrial service economy. The puzzle is what relationship this economy might have to the residual manufacturing base. This is closely related to another consideration: the paradox that, in one of the few still dynamic parts of the British economy, expanding job opportunities go side by side with substantial unemployment.

The chapter goes on to speculate on the future. Major investments in transport infrastructure during the 1980s and early 1990s – the M25 London orbital motorway, the third London airport at Stansted, the Channel Tunnel – could create new patterns of accessibility and thus new opportunities for growth. Some of the most important such opportunities may occur north-east of London, on the M11 Corridor from London to Cambridge, which already has seen a burgeoning of high tech industry, the so-called Cambridge Phenomenon. It could be that here is the source of yet another industrial revolution, which is not isolated or unique, but is part of a much wider pattern of development.

Finally, however, a comparison between M4 and M11 raises the question: what is the role, and what might be the role, of conscious strategic planning? The most startling conclusion of this study is how accidental the whole process of high tech growth has been: it arose from separate decisions by public and private actors in different compartments, often working in ignorance of each other's activities and indeed in contradiction to what planners thought they were doing. And it could happen again. What might happen, however, if they worked more closely, and consciously, together as they do in Britain's leading competitor nations? This conscious co-ordination between the two sectors, employing new forms of strategic planning, could signal a departure for advanced developments, in new locations, in the 1990s.

2

Defining high tech

What is 'high technology' industry? According to advertisements in newspapers and magazines, 'high technology' embraces forklift trucks, cars, shoes, most new industrial buildings, and even a Japanese Shinto shrine, which incorporates an electronic sound-sensor. There is debate about the merits of 'high technology' methods of delivering babies. From a designer's point of view, 'high technology' is a style or design of furnishing and so on, imitative of or using industrial equipment (*Chambers twentieth century dictionary*).

However, the *Chambers* dictionary offers an alternative though equally nebulous concept of 'high tech', which is that of 'an advanced, sophisticated technology in specialist fields, e.g. electronics, involving high investment in R&D'. It is with this approach to 'high technology' that we are concerned here, and in particular with identifying those industries that are associated with the production of technologically advanced products. The criteria by which such industries are identified, however, are problematic. One possibility is to identify high tech industries according to standard definitions (e.g. by Minimum List Headings (MLH), which is the finest-level official industrial classification in the UK) so that the national or subnational analysis of employment and output changes can be undertaken. Before investigating this possibility, however, we should discuss why we are interested in such industries and then consider how others have attempted to define them.

There are several reasons why we are interested in identifying industries that produce new types of technologically advanced products. Policy makers have shown a particular interest in these high technology industries as they are often perceived to be the dynamic job and/or wealth generating segments of the economy, with worldwide markets and high growth rates. A cursory look at the emphasis on 'high technology' industries in the literature of development agencies throughout the United Kingdom further shows the importance attached to these industries. From a planning viewpoint, the Department of the Environment (GB DOE 1983) has criticized local authority behaviour towards high tech industries, but does not adequately define what these industries are. In a later circular (GB DOE 1984) it is argued that high technology firms often occupy laboratory-type premises rather than traditional 'smokestack' factories and so should not be confined to traditional industrial locations. The circular sets out guidelines stating that 'rules of thumb' (e.g. concerning the percentage of workforce in different functions) should not be used in judging planning applications, but rather the prime function of the establishment should determine the applicability of the Use Classes Order (the official activity distinctions used for planning purposes). It would seem, then, that if the main function is manufacturing, then that is the appropriate Use type even if the majority of the workforce is in R&D or office jobs. With such emphasis given to 'high technology' industries it is important to provide some common definition or consistent nomenclature for them.

Several strands of literature form possible bases for increasing our understanding of high tech industries (see, for instance, Hall & Markusen 1983, Ellin & Gillespie 1983, Langridge 1984). The first strand concerns Kondratieff or long waves of economic change (see, for instance, Kondratieff 1935, Freeman *et al*. 1982, Perez 1983). Kondratieff did not, however, suggest an explanation for these 47- to 60-year-long economic cycles, although he did argue that depressed conditions served to stimulate human inventiveness, so leading to discoveries that were exploited during the rise of the next wave (Warren 1982).

In a second, related strand of research Schumpeter (1939) suggested that the cyclical pattern of growth in capitalist economies was due largely to the diffusion of major new technologies (see Freeman 1984). A third theoretical field is concerned with the product life cycle model, first developed by Kuznets and Burns in the 1930s and subsequently taken up by Vernon (1966) and Hirsch (1967), and the profit cycle theory (Markusen 1983); both highlight the importance of the early stages of product development. These three perspectives emphasize the importance of innovation and product development over time, and high tech industries are often taken to be those at the 'leading edge' of such developments (see also Rothwell & Zegveld 1981 and Gershuny & Miles 1983 on the growing importance of microelectronics, information technology and biotechnology).

Defining 'high technology' industries

While some theoretical appreciation of the nature of industrial innovation, and hence of the development of high tech industries, is crucial to a study of this kind (Harris & McArthur 1985), it is inevitable, if scale and location are to be assessed, that some operational definition of the activity will be required. When considering such an operational definition, a useful initial distinction is that between high tech products and high tech processes; that is, between those industries or companies making high tech products and those using high tech products in their production processes. If we consider the 'low tech' equivalents as well, this distinction produces a four-part matrix, as shown in Figure 2.1, by which we might classify industries or companies.

At one extreme in this classification, we can consider cases in which the assembly of high tech products such as computers may often be labour-intensive and based on 'low-technology' production processes. On the other hand, traditional industries may utilize extremely advanced production processes to make traditional low tech products. An example is the automobile industry, which accounted for 40% of all industrial robots deployed in the UK in 1981 (Brady & Liff 1983). The other two parts of the matrix cover, of course, those sectors producing high tech goods with high tech products, as seems to be the case in the aerospace industry, for instance, or low tech goods using low tech production processes. In reality, of course, these quadrants may merge, with a combination of low and high technology products being used in any production process for making any 'high' or 'low tech' product.

For the UK as a whole, Northcott and Rogers (1984) found that by 1983, 18% of all production processes contained microelectronics, with the highest levels being in factories in the paper, print and publishing and the food, drink and tobacco industries. The use of microelectronics in products was, however, restricted primarily to a few sectors (especially mechanical and electrical engineering). Hence a definition

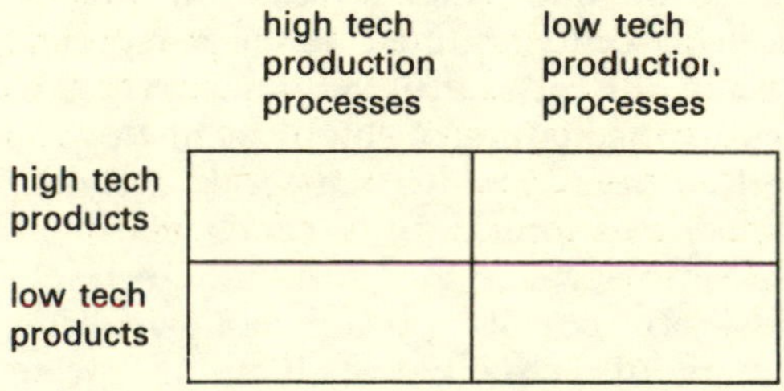

	high tech production processes	low tech production processes
high tech products		
low tech products		

Figure 2.1 High and low tech product and process categories.

based on aggregate industries which included high tech process users would incorporate many industries making traditional products. We wish to exclude such industries, although this is not to discount the importance of high tech user sectors, as these manufacturing and services industries are experiencing some of the greatest impacts of new technology (Braun & Senker 1982). One interesting way of using the matrix is as a basis for assessing how different regions are faring in the development and use of high tech products. Are industries in some regions predominantly in high/high category, or lagging behind in the low/low group? Some may be doing well in the introduction of high tech processes, but less well in the actual production of high tech goods. Goddard *et al*. (1985) found this to be the case with the depressed regions of the UK.

While a distinction between producers and users may be useful where firms can be unambiguously defined within such a framework, there are cases where such a distinction may be difficult to operate. For instance, how should computer software consultants be classified? They are not 'producers', in the strictest sense of the term, but neither are they merely 'users'. There is clearly a continuum from those using high technology products (e.g. insurance offices, manufacturers using robotics) to those purely developing such products, with various combinations in between (see Brook 1983). The concern here is with those industries at the producing end; that is, with high technology producers.[1]

One approach to defining high tech producers favoured by the Advisory Council for Applied Research and Development (GB ACARD 1979), the National Economic Development Council (GB NEDC 1983), and the Manpower Services Commission (Brady & Liff 1983) is to speak of 'new' technology. This has the attraction of avoiding the issue of deciding what constitutes 'high' technology by classifying technologies in terms of temporal rather than qualitative considerations. Such an approach seems to have operational simplicity; that is, industrial processes can be classified according to the period in which the crucial technology was developed.

However, this would still leave unresolved the problem of classifying industries and firms, particularly where the technologies involved have emerged over a period of time which straddles a number of phases of industrial innovation. Also, to assume that 'high' technology and 'new' technology are necessarily one and the same requires considerable value judgement as to the nature of 'high' technology. Yet, although this approach might be operationally manageable, it is open to abuse. ACARD, for instance, has produced a series of reports on specific technologies in which not only is the term 'new technology' used in a purely arbitrary fashion, but it appears alongside the terms 'high technology' and 'emergent technology' as if all three were interchangeable (see, for instance, GB ACARD 1979).

Another problem with this approach is the need to 'date' the introduction of 'new' technology. In practice this is likely to vary across industries, regions and nations. For instance, the silicon chip was developed in Britain (by Plessey) and in the USA for military and space purposes, yet its application in the UK to personal computers took place some years later. So even with a definition of high tech restricted to electronic computers, problems encountered in 'dating' high tech are likely to be considerable.

Clearly, then, there are both conceptual and operational difficulties with a purely product-based definition. However, there is a far more fundamental issue to consider, namely, the degree of subjectivity inherent in such a method of identification. Industries, firms and activities are usually identified as high tech neither because they exhibit particular qualities, nor because they perform in a particular way. Rather, they are identified on the basis of value judgements of what 'high technology' should be. As Breheny *et al.* (1983) point out, the approach of making value judgements on what industries are to be included as 'high technology' (for instance, Oakey 1981, SCLSERP 1983) gives a working definition, but has no objective base from which to make comparisons.

A second approach to identifying high tech producer industries is to specify certain key characteristics and to use some objective measure of these to identify the industries. For example, Rogers and Larson (1984, p. 29) provide a set of characteristics for high tech industries; they have highly skilled employees, many of whom are scientists and engineers; a fast rate of growth; a high ratio of R&D expenditure to sales; and a worldwide market for their products. In order to develop an operational definition, however, these characteristics have to be specified more precisely.

Using British data, Langridge (1984) has tested a number of possible variables, including employment growth rates, R&D expenditure as a percentage of sales, and the ratio of administrative, technical and clerical staff to operatives. He found that for each characteristic used, a number of anomalies arose in which industries that few would call high tech (according to some subjective or product basis) were included, and commonly accepted high tech industries excluded. This was because the data were poor proxies; or they represented only one aspect of high tech industries; or the chosen measure was not necessarily a high tech characteristic (e.g. employment growth).

Ellin and Gillespie (1983) used first Engineering Industry Training Board (EITB) occupational data for the engineering industries to calculate the percentage of total employment accounted for by scientists and technicians and secondly R&D expenditure as a share of net output, in order to identify several high tech industries (by Minimum List Headings) which were highly ranked in each of the two cases (see Table 2.1). The same engineering industries were ranked top in each case. However, the problem with this approach was the restricted set of industries used, due to data limitations.

Defining high tech

Table 2.1 British definitions of high technology industry.

MLH	SCLSERP (1983)	GB Electronics EDC(1973)	Ellin & Gillespie (1983)	This study
general chemicals	271			
pharmaceutical chemicals & preparations	272		272	272
synthetic resins, plastics & rubber	276			
office machinery	338			
scientific & industrial equipment	354	354		
telegraph & telephone apparatus	363	363	363	363
radio & electronic components	364	364	364	364
broadcast receiving & sound equipment	365	365	365	365
electronic computers	366	366	366	366
radio, radar & electronic capital goods	367	367	367	367
aerospace equipment manufacturing	383	383	383	383
research & development services	876			

Note: Based on 1968 SIC.

In a US study, Hall and Markusen (1983) considered three criteria for identifying high tech industries, together with some of the problems of using such criteria (see also Glasmeier *et al.* 1983). First, they looked at the degree of technical sophistication of the product process; an approach which has the problems of subjectivity discussed above. Secondly, growth in employment was used, but this has the problem that it may reflect demand and capital intensity rather than advanced technology. Thirdly, R&D expenditure as a percentage of sales was investigated. This proved both difficult to measure and to have a bias towards industries in early stages of product development.

Some of these characteristics had been used by other writers such as Rees (1979), who also assessed output growth rates. Premus (1982) expanded the relevant characteristics to include: labour intensity in production processes; high proportions of technicians, scientists and engineers; a reliance on the application of advances in science; and the importance of R&D inputs. He included five two-digit SIC (Standard Industrial Classifications) industries in his definition (see Table 2.2). The same five two-digit SIC industries were arrived at in a definition devised by Boretsky (1982) and used by the US Department of

Commerce (1983). These five industries are described as 'technology intensive', with selected three-digit industries within the two-digit level being labelled as 'high technology' industries. The criteria used in both cases were the proportion of gross product of an industry spent on R&D and the proportion of total employment consisting of scientists, engineers and technicians. The cut-off points for the two definitions were:

As a general proposition *technology intensive* industries are defined as those which normally spend 5 per cent or more of their gross product (BEA concept of value added) on R and D and/or normally 5 per cent or more of their total employment consists of 'natural' scientists, engineers and technicians. *High technology* industries normally spend at least 10 per cent of their gross

Table 2.2 United States definition of high technology industry.

Rank	SIC	Title
1	376	missiles
2	357	office computing machines
3	381	engineering, laboratory instruments and scientific instruments
4	366	communication equipment
5	383	optical instruments and lenses
6	286	industrial organic chemicals
7	372	aircraft and parts
8	283	drugs
9	291	petroleum refining
10	382	measuring and controlling instruments
11	367	electronic components and assembly
12	281	industrial inorganic chemicals
13	282	plastics and synthetic resins
14	351	engines and turbines
15	348	ordnance
16	289	miscellaneous chemicals
17	386	photographic equipment
18	362	electrical industrial apparatus
19	361	electrical transmission equipment
20	353	construction equipment
21	285	paints
22	303	reclaimed rubber
23	356	general industry machinery
24	374	railroads
25	365	radio and TV receiving equipment
26	267	agricultural chemicals
27	354	metalworking machinery
28	364	medical and dental supply
29	284	soap

Source: Hall and Markusen (1983).

product (value added) on R and D and/or at least 10 per cent of their total employment consists of 'natural scientists, engineers and technicians'. (US Department of Commerce 1983, p. 35)

Hall and Markusen (1983) conclude their argument for a precise, objective and fully comparable definition by proposing the use of a single criterion: the percentage of an industry's labour force that is in technical occupations. This provides a standard measure of that industry's capacity to employ scientific and technological practices and to generate scientifically and technologically advanced products. The advantages of this measure are that it is simple to apply, it is based on a standard classification of occupations (so is comparable across industries and avoids the measurement problems of other variables), the data are easily available, and it avoids the problems of weighting a composite index of various characteristics. Hall and Markusen (1983) go on to identify 29 three-digit SIC industries in the USA with higher than the national manufacturing average of engineers, engineering technicians, computer scientists, scientists and mathematicians. Table 2.2 lists these 29 industries ranked according to this criterion.

Identifying British 'high technology' industries

In attempting a definition of UK high technology industries, the M4 study initially adopted the Hall and Markusen (1983) approach and sought to identify industries with a share of engineers, technologists and scientists above the average for British manufacturing industries. This gives 36 (Minimum List Headings on the 1968 base) manufacturing industries and 0 service industries (Table 2.3). This 'first sieve' definition is, then, reasonably comparable to the Hall and Markusen (1983) three-digit definition. However, in addition, a 'core' set of seven manufacturing industries which have relatively high R&D expenditure are isolated. This 'core' or 'second sieve' group (see Table 2.3) is therefore comprised of industries with labour forces reflecting their capacity to produce advanced products, and also relatively high shares of material and personnel resources devoted to producing advanced products.

Unfortunately there are some data problems with this 'core' classification. R&D expenditure data were unavailable for service industries, and hence the second sieve exercise could not be carried out for this sector. Otherwise, at least research and development services would be selected. Although the analysis here is restricted to manufacturing high tech sectors, it would have been interesting to see how the service sectors fare when the two sets of criteria are applied. Another problem is that individual R&D data were not available for SIC 363, 365 or 367. Broadcasting equipment (SIC 365) might otherwise

Table 2.3 'First sieve' definition of UK high technology industries.

MLH		% R&D of value added in selected industries 1978	% Occupation 1971
383	aerospace equipment manufacturing	39.59	12.45
366	electronic computers	35.71	22.13
367	radio, radar & electronic capital goods	n.a.	15.14
364	radio & electronic components	23.18	10.62
363	telegraph & telephone apparatus	n.a.	7.57
365	broadcast receiving & sound equipment	n.a.	4.30
272	pharmaceutical chemicals & preparations	20.62	7.53
279	other chemical industries	6.18	3.53
276	synthetic resins, plastics & rubber	5.98	8.32
262	mineral oil refining	5.83	10.45
369	other electrical goods	4.84	4.28
361	electrical machinery	3.91	7.31
274	paint	3.89	5.46
362	insulated wires & cables	3.64	5.42
380	wheeled tractor manufacturing	3.41	4.22
354	scientific & industrial instruments & equipment	3.30	9.35
351	photographic & document copying equipment		8.41
335	textile machinery & accessories	3.24	4.19
339	other machinery	2.41	5.05
332	metal working machine tools	1.78	4.49
337	mechanical handling equipment	1.22	4.53
411	production of man-made fibres	1.05	4.72
341	industrial plant & steelwork	0.90	6.91
271	general chemicals	n.a.	10.17
277	dyestuffs & pigments	n.a.	9.93
338	office machinery	n.a.	7.62
334	industrial engines	n.a.	7.38
263	lubricating oils & greases	n.a.	7.18
278	fertilizers	n.a.	6.81
342	ordnance & small arms	n.a.	5.68
275	soap & detergents	n.a.	5.48
333	pumps, valves & compressors	n.a.	5.46
384	locomotive & railway track equipment	n.a.	5.46
323	other base metals	n.a.	4.84
311	iron & steel (general)	n.a.	4.13
336	construction & earth moving equipment	n.a.	3.85

continued

Table 2.3 *continued*

MLH		% R&D of value added in selected industries 1978	% Occupation 1971
	Service industries		
876	research & development services	n.a.	29.99
879.2	other scientific & technical services		17.99
879.1	architects, surveyors & consulting engineers	n.a.	12.21
879	other professional & scientific services	n.a.	9.83
602	electricity	n.a.	8.74
601	gas		5.77
866	central offices not available elsewhere	n.a.	5.70
879.3	professional & scientific organizations	n.a.	5.40
707	air transport	n.a.	3.79
865	other business services	n.a.	3.75

Sources: Business Monitor M014, 1978, HMSO; Census, 1971, *Economic Activity* Part III, Table 19, HMSO.
n.a., not available.

fail to satisfy the criteria. The data on R&D expenditure also has possible problems of under-measurement for small firms, where management may undertake informal and unrecorded R&D, thus leading to an underestimate of R&D expenditure in the industries with a high proportion of such firms (for example, scientific instruments; see Freeman *et al*. 1982). Also, the R&D links between defence and industry are well known (Gummett 1984), with over half of the government's R&D expenditure going on defence through private firms (GB Cabinet Office 1984). Clearly, the R&D data in Table 2.3 will reflect the high levels of recorded R&D in certain defence-dominated industries, and those industries producing advanced technology products for predominantly non-defence markets may be under-estimated.

As most current British studies of high technology industries are restricted to data using 1968 Standard Industrial Classifications for pre-1981 time-series, the most reliable source with the necessary occupational and industry data is the 1971 Census. The definition can be updated and based on the 1980 SIC when data become available on R&D expenditure for 1980 SIC industries. An attempt has been made here to produce a 1980 SIC equivalent of the 1968 SIC definition. However, because the data required to repeat the 1968-based definition are not available, a simple cross-referencing approach was used, in which 1980 SIC industries corresponding to the seven MLHs making up our 1968 definition were selected. The correspondence between the two SIC classifications was based on a cross-reference list provided by the Department of Employment. The industries included in the resulting 1980 definition are shown in Table 2.4. Under the 1980 classification the finest categories are referred to as 'Activity Headings' rather than 'Minimum List Headings'. This method of arriving at a 1980 SIC definition of high tech industries is far from satisfactory, but is necessary if we are to carry out any post-1981 analysis. There are well-known problems with the 1968 SICs, although the 1980 version (and the occupational classification used in 1981) takes account of some of the major changes in industrial activity (e.g. computer services, which are included under 'other business services' in the 1968 classification, possess their own category in the 1980 SIC). However, when considering local employment changes over time there is currently little alternative to using the 1968 SIC.

Using such broad categories as Minimum List Headings (or Activity Headings for 1980 SICs) has the usual problems of heterogeneity within the industry, changes in the industry composition over time, and collection problems (e.g. the head office of an international computer firm may be listed under 'central offices'). An example of the first two problems is MLH 364, radio and electronic components, which contains firms engaged both in the production of radio valves (not usually considered 'high technology') and increasingly in the production of solid state circuitry. Conversely, biotechnology is still too small a part of the relevant MLH to be counted as high tech, and parts of scientific and industrial instruments (which fall outside our definition) undoubtedly have the relevant high tech characteristics.

If a number of high technology industries (or MLHs) are added together, then this problem of aggregation is compounded. Although each of the industries shares a common characteristic such as occupational composition, they may have very different spatial distributions or employment behaviour. Even within an industry there

Table 2.4 1980 SIC definition of UK high technology industries.

Activity	Heading
2570	pharmaceutical products
3302	electronic data processing equipment
3441	telegraph & telephone apparatus & equipment
3442	electrical instruments & control systems
3443	radio & electronic capital goods
3444	components other than active ones, mainly for electronic equipment
3453	active components & electronic subassembly
3454	electronic consumer goods & other electronic equipment
3640	aerospace equipment manufacture & repair
7902	telecommunications (excluding PO, broadcasting & cable)

may, of course, be significant variations in characteristics across regions (Breheny & McQuaid 1984). Hence it is preferable to analyse each industry (i.e. broad product group) separately, where possible.

A further approach to definition that should be mentioned is that which identifies high technology firms or establishments rather than industries (see, for example, Firn & Roberts 1984 on Scottish high tech firms; and Brook 1982). Arthur D. Little (1977, p. 2) defines 'new technology based firms' as those based on a patented invention or with substantial technological risk (as well as being new and set up by an individual or group of individuals to exploit a technological innovation).

A firm or establishment basis for defining high tech industries may be useful for policy purposes. Such a basis may overcome the problem of aggregation discussed above and may also be appropriate for studies wishing to include those firms using advanced production processes (possibly using some technology-based criterion). However, for studies of national or international trends and spatial distributions, or local studies using government data sources, such an approach is likely to be complementary rather than a substitute for an industry-based approach of the kinds discussed here.

Conclusion

This review of definitions of high technology industries shows that the approach adopted in this study corresponds reasonably well with a general consensus on the most appropriate criteria for an industry-based definition of high technology activity. Furthermore, the individual Minimum List Headings chosen match up closely to those selected in other British studies.

The method adopted here for identifying what are commonly called high technology or high technology producer industries provides an 'objective', simple measure which allows national and international comparisons. Conceptually, it provides a reasonable reflection of what is considered to be a key aspect of such industries: that is, the industry's capacity to employ scientific and technological practices and the generation of scientifically and technologically advanced products. The 'core' industries also have a high share of value-added composed of R&D. Such industries are not static and will vary over time as once-new innovations become routine (for instance, the automobile). The definition used here allows constant review and so can reflect such changes.

No apologies are made for concentrating on a clear operational definition which is consistent, exclusive and exhaustive, rather than taking a more abstract, theoretical view. A number of major data problems remain; particularly those concerned with the aggregate level of information, the fact that Minimum List Headings are constructed on a basis that does not necessarily reflect the level of product technology, and potential biases in R&D expenditure measurement. Hence, each industry contains some segments or firms which few would consider to be high tech, while excluding others that may well be. The considerable gap between the conceptualization of high tech and identifying those concepts in operational terms needs to be fully recognized.

Note

1   Others (e.g. Debenham *et al.* 1983) have suggested using the term 'knowledge-based' industries rather than 'high tech'; however, this is wider than our usage and also may lead to confusion with previous writers who have used the term 'knowledge economy' (e.g. Drucker 1969) or others such as Bell (1974, 1979) who have considered knowledge as a key resource within the wider framework of the post-industrial society.

PART II   Where?

3

The geography of high tech in Britain

The general enthusiasm for high technology industry in Britain – fed, for different reasons, by the media and the property industry – has created two popular myths about its potential for development. The first is that high tech has been the major national generator of new jobs in recent years. The second is that this growth has been concentrated in a few favoured locations, most notably in a continuous corridor of growth from London to Bristol and on into South Wales: Britain's 'M4 Corridor', equivalent of California's Silicon Valley. Both assumptions need critical appraisal.

This chapter begins the appraisal by addressing two specific issues: the role of high technology, as defined in the last chapter, in the national economy; and the geography of high technology industry in Britain, relative to other developments in the space economy. The next chapter continues the appraisal by investigating the nature of high technology in the county of Berkshire – arguably the core of the corridor, if corridor it be.

High technology industry and national employment change

In recent years the British economy has been characterized by declining levels of employment and dramatically increasing levels of unemployment. This general picture disguises contraction in the manufacturing sector (1 963 200 or 24.9% jobs lost 1971–81) and modest but significant gains in the service sector (461 984 or 10.0% jobs gained 1971–81). Such a shift to service employment reflects a trend in all industrialized nations, but the decline in manufacturing has been more marked in Britain than in most other industrial countries.

National employment changes in our short-listed, two-sieve, high technology group of industries are shown in Tables 3.1 and 3.2. Table 3.1 shows employment levels and changes on the 1968 SIC basis over the 1971–83 period; post-1981 changes are included in the hope of picking up what we assume to be relatively recent changes in these high technology sectors.

We can see immediately from Table 3.1 that at the national level the much vaunted 'high technology' group, upon which so much hope seems to be pinned, contributes not to employment gain, but to substantial loss. In fact, over 15% of these high technology jobs were lost over the period. Only one of the 1968 SIC sectors – radio, radar and electronic capital goods (367) – actually increased its employment level over the 1971–83 period. Electronic computers (366) shows some growth in recent years, but the general trend in the other sectors is a steady loss of jobs.

Table 3.2, which makes use of the more discriminating 1980 SIC, and which should thus give a finer measure of high technology job change, presents a picture that is consistent with Table 3.1. The percentage decline

Table 3.1 Employment change in aggregate high technology sectors, 1971–83, on 1968 SIC base.

MLH	June 1971	June 1975	Sept. 1981	Sept. 1983	Change 1971–81	Change 1975–83
272 pharmaceutical chemicals & pre-parations	72 600	76 100	69 300	71 800	− 3 300	− 4 300
363 telegraph & tele-phone apparatus	84 300	87 000	64 500	56 200	−19 800	−30 800
364 radio & electronic components	128 300	128 400	103 100	106 800	−25 200	−21 600
365 broadcast receiving & sound equipment	48 300	54 900	37 500	22 500	−10 800	−32 400
366 electronic computers	50 100	43 300	42 100	58 100	− 8 000	14 800
367 radio, radar & elec-tronic capital goods	94 100	89 300	100 100	105 300	6 000	16 000
383 aerospace equip-ment manufactur-ing & repair	211 400	204 400	197 300	163 000	−14 100	−41 400
Total	689 100	683 400	613 900	583 700	−75 200	−99 700

Sources: British Labour Statistics; Department of Employment Gazette.

Table 3.2 Employment change in aggregate high technology sectors, 1981–4, on 1980 SIC base.

Activity Heading	Dec. 1981	Sept. 1984	Change 1981–4
2570 pharmaceutical products	81 200	82 300	1 100
3302 electronic data processing equipment	72 200	74 700	2 500
3443 radio & electronic capital goods & components	200 500	205 100	4 600
3454 electrical & electronic consumer goods & components	125 300	135 700	10 400
3640 aerospace equipment manufacture & repair	182 100	157 800	−24 300
3441 telegraph & telephone apparatus & equipment	231 700	224 200	− 7 500
Total	893 000	879 800	−13 200

Source: Department of Employment Gazette.

in jobs, at 1.5% over the 1980–4 period, is not as dramatic as that under the 1968 SIC classification, but is nevertheless very disappointing. Only pharmaceuticals and, again, the electronics sectors show any growth.

The dismal record of high technology in Britain contrasts remarkably with its performance in the USA (Table 3.3). Markusen *et al.* (1986) estimate that between 1972 and 1981, using the single 'sieve' definition explained in the last chapter, the American economy gained over one million high technology jobs – 728 000 of them in the brief period 1977–81 alone. A comparable single 'sieve' definition for Britain, as explained in Chapter 2, gives 43 MLHs which accounted for 3 619 800 jobs in Great Britain in 1971 and 3 253 200 jobs in 1981; a decline of 366 600 'high technology' jobs (10%), 204 000 of which occurred between 1975 and 1981. Table 3.3 also shows a comparison of the UK double 'sieve' definition and the equivalent US three-digit sectors over the 1971–81 and 1972–81 periods respectively. The seven UK sectors contributed a loss of 75 200 (−10.9%) jobs, whereas the ten US sectors were responsible for a gain of 631 300 (+29.9%).

The geography of job gains and losses

Two related, overlapping trends dominate the changing economic geography of Britain in the 1970s and early 1980s. One is the continuing prosperity of the south of England relative to other regions – an advantage reinforced, at the moment, by relative success in the service and high technology sectors. The other is the general

Table 3.3 Great Britain and the United States: high tech employment.

Great Britain				United States			
	1981	1971–81			1981	1972–81	
	000s	000s	%		000s	000s	%
pharmaceutical chemicals & preparations	69.3	−3.3	−4.6	pharmaceutical preparations	130.5	+18.5	+16.5
telegraph & telephone apparatus	64.5	−19.8	−23.5	telephone & telegraph apparatus	147.4	+13.0	9.7
radio & electronic components	103.1	−25.2	−19.6	radio & TV transmission etc.	426.9	+107.7	+33.7
broadcast receiving & sound equipment	37.5	−10.8	−22.4	electronic computing equipment	320.7	+175.9	+121.5
				aircraft	301.1	+69.3	+29.9
electronic computers	42.1	−8.0	−16.0	electronic components, NEC	190.0	+89.5	+89.1
radio, radar & electronic capital goods	100.1	+6.0	+6.4	semiconductors & related devices	169.5	+71.9	+73.7
aerospace equipment manufacturing	197.3	−14.1	−6.7	construction machinery equipment	145.9	+12.1	+9.0
				aircraft parts & equipment, NEC	140.3	+38.1	+37.3
				aircraft engines & parts	140.0	+35.3	+33.9
Total (7)	613.9	−75.2	−10.9	Total (10)	2112.3	+631.3	+29.9
Total (43)	3253.2	−366.6	−10.1	Total (100)	5475.2	+1080.7	+24.6

Sources: Breheny & McQuaid 1985, based on GB Annual Census of Employment; Markusen *et al.* 1986, based on US Annual Survey of Manufacturers, 1981, and Census of Manufacturers, 1972 and 1977.

movement of economic activity away from major urban areas and out towards suburban or exurban areas.

Table 3.4, which shows employment change in the South-East during 1971–81, reflects both these trends. At first glance, the South-. East has not done particularly well relative to Great Britain as a whole. Its manufacturing job loss (−23.9% in 1971–81) is only marginally better than the nation (−24.9%), and its service sector employment growth (+10.00%) is poorer than that for Great Britain (+14.5%). Overall the South-East's employment change was only slightly more

Table 3.4 Employment change, south-east England, 1971–81.

	Manufacturing		Services		Total	
	Abs.	%	Abs.	%	Abs.	%
Inner London	−186 722	−40.6	−105 041	−6.1	−318 300	−13.9
Outer London	−192 047	−32.6	98 681	10.2	−95 662	−5.8
OMA	−125 370	−18.4	237 233	25.0	111 604	6.3
OSE	−15 337	−3.5	231 111	23.9	210 487	13.6
South-East	−519 476	−23.9	461 984	10.0	−91 871	−1.3
Great Britain	−1 963 200	−24.9	1 703 000	14.5	−501 200	−2.3
Berks, Bucks, Hants, West Sussex	−11 106	−3.0	208 190	32.1	196 595	17.6
Other ROSE counties	−129 601	−17.2	260 154	20.5	125 496	5.7

Source: SERPLAN (1985) (employees in employment, 1968 SIC).

favourable than that for Great Britain (−1.3% compared to −2.3%). However, the table makes it clear that the South-East's overall showing is heavily weighted by London's, and particularly Inner London's, poor performance. The Outer Metropolitan Area (OMA) and the Outer South-East (OSE) perform considerably better on all counts than Great Britain or the South-East as a whole.

Table 3.4 in fact demonstrates very clearly the trend in the movement of economic activity away from major cities. Apart from the performance of the OMA in service employment, the percentage change figures show a neat and consistent pattern of high levels of decline nearer the urban core and increasingly better performance away from it. Figure 3.1 demonstrates this pattern of change very clearly. Towns and villages in the Western Corridor, then, perform well because places in these outer southeastern rings do well – whether north, south, east or west.

However, these rings of prosperity are not homogeneous: though they do show consistently higher growth away from London, there is also a distinct western bias. As Table 3.4 shows, the 1971–81 performance of the four western Rest of South-East (ROSE) counties – a decline of only 3.0% in manufacturing, a rise of 32.1% in services and an overall growth of 17.6% – is considerably better than that the overall ROSE record during the period.

In the 'Corridor' counties outside the South-East, we find that the pattern of manufacturing decline and service growth is repeated consistently. Table 3.5 shows employment change in the two broad sectors. South Glamorgan has fared worst because of a high manufacturing loss and a low service gain. The most dramatic restructuring has taken place in West Glamorgan, where a 36.5%

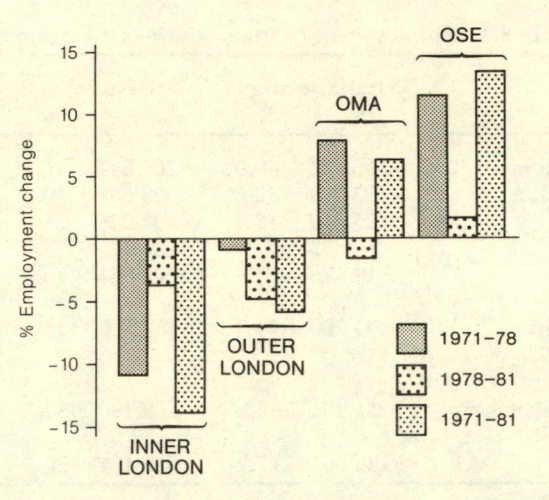

Figure 3.1 Employment change in south-east England.
Source: SERPLAN (1985).

Table 3.5 Employment change in non-south-east 'Corridor' counties, 1971–81.

	Manufacturing		Services		Total	
	Abs.	%	Abs.	%	Abs.	%
Wiltshire	−13 956	(−21.3)	29 804	(32.7)	16 442	(9.4)
Avon	−25 264	(−20.1)	29 914	(13.6)	2 798	(0.7)
Gwent	−19 763	(−27.5)	8 869	(12.4)	−11 767	(−7.3)
South Glamorgan	−16 896	(−39.8)	3 531	(2.9)	−16 683	(−9.3)
Mid Glamorgan	−10 656	(−16.3)	12 227	(14.6)	−8 331	(−4.9)
West Glamorgan	−22 872	(−36.5)	21 977	(32.5)	−9 948	(−6.5)

Source: Annual Census of Employment (employees in employment, 1968 SIC; manufacturing = SICs 3–19, services = SICs 21–7).

decline in manufacturing has been countered by a 32.5% increase in the service sector.

This schematic analysis of recent employment changes – nationally, regionally, and in the counties of the Corridor – provides a necessary framework within which we can better understand the contribution of high tech manufacturing. Particularly, it should caution us against undue optimism about that contribution. Manufacturing in Britain is in deep decline; and, contrary to popular opinion, high tech is part of that decline. The question is whether this is even true of the Corridor itself.

Identifying high tech
geographical concentrations

Given the intense interest in the M4 Corridor, and in other supposed pockets of high technology industry in central Scotland and around Cambridge, remarkably few efforts have been made to measure the extent of these concentrations. For this, data problems are largely responsible.

Gibbs (1983) and Ellin and Gillespie (1983) have produced pictures of high tech concentrations in Britain using 1971 and 1978 ACE (Annual Census of Employment) statistics for functional regions. They use a particular grouping of MLHs to define high technology industry, and plot levels and changes in aggregate employment in these industry groups. These studies conclude that while the South-East generally shows the highest concentrations of high technology industry, there is no identifiable M4 Corridor (Gibbs 1983, p. 21). One of the problems with this aggregate approach, also characteristic of similar American work (e.g. Glasmeier 1985), is that it disguises important locational differences between one high technology industry and another. Also, the use of 1978 and even 1981 data to provide a 'current' picture of the location of high technology industry may be misleading; important developments may well have occurred since 1981.

An Engineering Industry Training Board (EITB) report of 1984 has produced interesting results from a survey of its members in the electronics industry. Employment is concentrated in the metropolitan counties of London, West Midlands, Manchester and Strathclyde, plus the southeastern counties of Berkshire, Hampshire, Essex and Hertfordshire. Employment change in electronics over the period 1978–83 presents a different picture. None of the metropolitan counties grew by more than 200 jobs. Berkshire is most prominent in absolute job gain, with over 4000 new jobs; Gwent and Lothian gained between 1000 and 4000 (EITB 1984). But, as the report says, it is difficult to identify an M4 Corridor, at least in electronics.

Howells's (1984) study of the location of R&D employment (not one of our high technology manufacturing sectors, but a likely 'high technology' representative in the service sector) identifies significant concentrations in the South-East. In particular, it shows R&D employment in Berkshire and Hampshire to have risen dramatically over the 1971–6 period. This would seem to give some support to the concept of an M4 Corridor, if only at its eastern end. However, the Annual Census of Employment (ACE) data used by Howells show wild fluctuations over the 1971–81 period, indicating that they are suspect. This instance highlights a major problem in using ACE data at fine geographical and sectoral levels. The data seem to include considerable errors and reclassifications, including the treatment of headquarters office employment, which make cross-sectional and time-series analysis very difficult.

In absolute terms, employment in our seven short-listed high technology industries, in 1981, was concentrated in the large metropolitan areas of London (over 91 000 jobs), the West Midlands (36 500 jobs) and Greater Manchester (27 300 jobs), and in some surrounding counties, especially Hertfordshire (45 000 jobs) and Hampshire (33 800 jobs) around London, and Lancashire (30 000 jobs) north of Greater Manchester. The dominance of the metropolitan areas in absolute high technology employment reflects, of course, their size as employment centres rather than any particular concentration of high technology activity. It also reflects the inadequacy of our definition of high technology industries, because clearly we are including companies classified according to our MLHs but which are not involved in the production of new technology goods. The distribution of high technology jobs in Great Britain at 1981 is shown in Figure 3.2 for all counties with more than 15 000 such jobs.

A useful way of measuring the relative significance of concentrations of jobs, which allows for the different size of the employment base in different areas, is the location quotient. This relates the proportion of local jobs accounted for by a particular sector to the proportion of national jobs accounted for by that same sector; the ratio between the two is the quotient. A quotient higher than unity means that the sector is concentrated in that locality. Table 3.6 lists those counties with

Table 3.6 Location quotients (above 1.0) for counties with 5000 or more high technology jobs, 1981.

County	Location quotient
Hertfordshire	3.60
Avon	2.32
Fife	2.24
Somerset	2.11
Berkshire	2.08
West Sussex	2.03
Lancashire	2.01
Hampshire	2.00
Essex	1.83
Derbyshire	1.83
Surrey	1.79
Mid Glamorgan	1.76
Dorset	1.69
Clwyd	1.49
Nottinghamshire	1.32
Gloucestershire	1.24
Bedfordshire	1.24
Buckinghamshire	1.15
West Midlands	1.02

Note: Ranked by location quotient; base = GB.

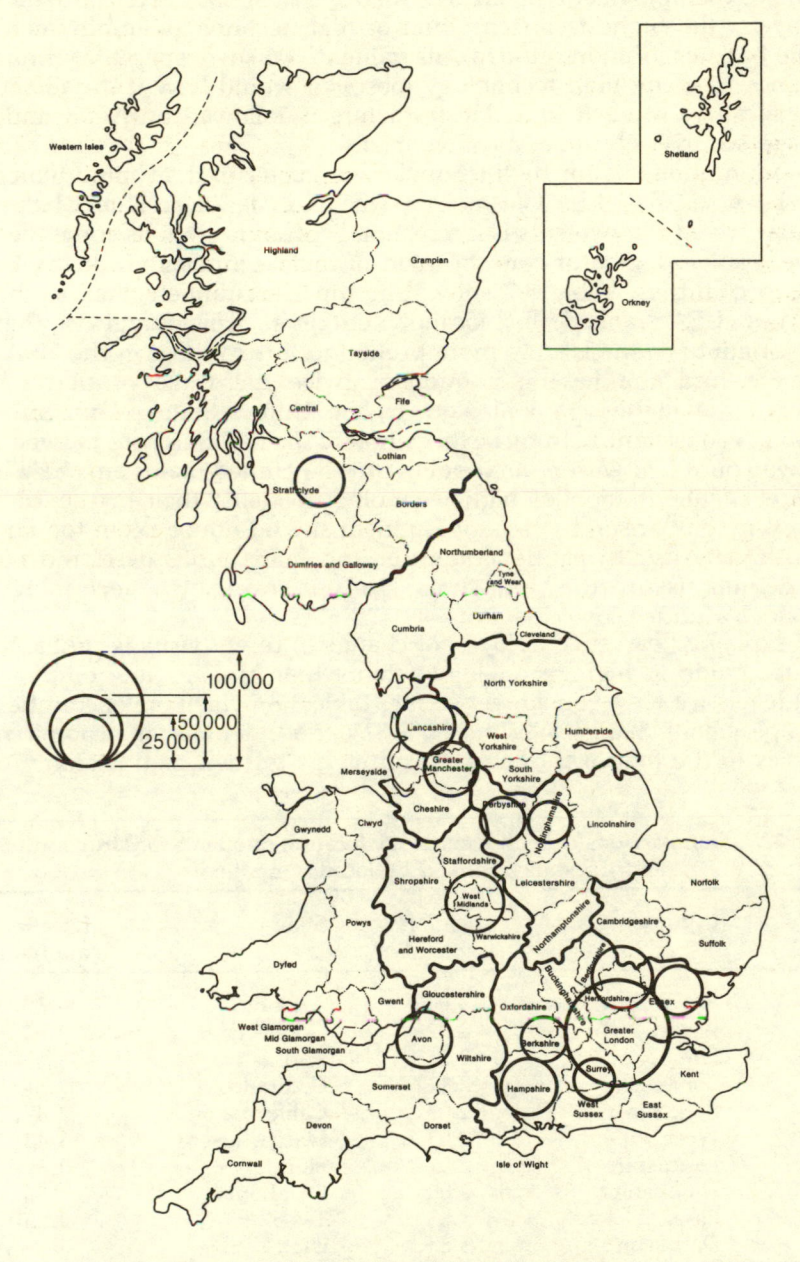

Figure 3.2 Location of high technology employment, 1981 (counties with 15 000 or more jobs). *Source*: Annual Census of Employment.

aggregate high technology location quotients of more than 2.0 and high technology employment of greater than 5 000 in 1981. Hertfordshire has by far the highest concentration of high technology employment on the basis of location quotient, its value of 3.6 suggesting that it has 3.6 times as many high technology jobs as it would have if it had the national share of such jobs. Hertfordshire is followed by Avon and, perhaps surprisingly, Fife and Somerset.

Location quotients for high technology industry in the United States have been calculated by Glasmeier (1985) and Markusen *et al.* (1986). Comparison of the two sets is interesting because it gives us some idea of the relative degree of concentration of high technology industry in the two countries. Table 3.7 shows the top ten counties/states in the UK and USA ranked by location quotient. This suggests that concentrations in the UK are more pronounced than those in the USA, with Hertfordshire having a quotient twice the value of that for Arizona, the highest ranked state in the United States. This comparison is consistent with our expectation of likely differences between the two countries, with federal politics in the United States ensuring a more even distribution of high technology benefits than in the UK. However, comparison by location quotient should not be taken too far, because values will vary depending on the spatial units used and on the definitions of the industries – in our case high technology industries – under consideration.

As explained earlier, employment change in recent years is probably a better guide to high technology activity than the absolute employment levels we have considered so far. Table 3.8, which ranks counties by employment change over the 1975–81 period, demotes metropolitan counties to the foot of the table in contrast with their high ranking in

Table 3.7 Comparisons of high technology location quotients for UK counties and US states, ranked by location quotient.

Rank	County	Location quotient	State	Location quotient
1	Hertfordshire	3.60	Arizona	1.80
2	Avon	2.32	Connecticut	1.65
3	Fife	2.24	Kansas	1.63
4	Somerset	2.11	Colorado	1.57
5	Berkshire	2.08	California	1.49
6	West Sussex	2.03	Massachusetts	1.43
7	Lancashire	2.01	Florida	1.39
8	Hampshire	2.00	Oklahoma	1.38
9	Essex	1.83	Texas	1.33
10	Derbyshire	1.83	Utah	1.33

Source: US data taken from Glasmeier (1985), p. 63.
Note: UK 1981 base, USA 1977 base.

Table 3.8 Aggregate high technology employment 1981, change 1975–81, location quotients 1981, by selected counties, ranked by 1975–81 change.

Rank	County	Absolute employment 1981	Change in employment 1975–81	Percentage change 1975–81	Location quotient 1981
1	Berkshire	19 732	7 600	62.2	2.08
2	Hertfordshire	45 060	5 914	15.1	3.60
3	Clwyd	5 094	3 283	181.3	1.49
4	Hampshire	33 807	2 179	6.8	2.00
5	Surrey	18 191	1 811	11.1	1.79
6	Kent	14 650	1 679	12.9	0.96
7	West Sussex	14 694	1 669	12.8	2.03
8	Bedfordshire	7 264	1 653	29.5	1.24
9	Lothian	9 042	1 461	19.3	0.93
10	Lancashire	30 033	1 392	4.8	2.01
58	Strathclyde	23 575	−3 847	−14.0	0.88
59	Tyne and Wear	8 490	−4 032	−32.2	0.62
60	Essex	24 649	−4 342	−14.9	1.83
61	Nottinghamshire	15 860	−5 123	−24.4	1.32
62	Merseyside	14 417	−6 435	−30.8	0.86
63	West Midlands	36 500	−8 257	−18.4	1.02
64	Greater London	91 448	−17 012	−15.7	0.85
	Great Britain	640 874	−42 429	−6.2	1.00

Source: Annual Census of Employment.

absolute, point-of-time, high technology employment terms. The counties of Berkshire, Hampshire and Hertfordshire come high in the rankings in this analysis of change. Clwyd shows up well, but there is a suspicion that there is some statistical anomaly involved that gives the county more credit than it is due. The mapping of the employment change data in Figure 3.3 demonstrates the dominance of the top-ranking counties.

Disaggregating the pattern

Such aggregate analysis of high technology industry has one major weakness: it disguises significant differences between one high tech industry and another. To illuminate these differences, Table 3.9 ranks counties by location quotients (above 2.0) for each of our seven high technology sectors, and quickly gives an impression of those counties with high relative concentrations in each MLH. For instance, aerospace (MLH 383) is concentrated in a few counties, such as Avon, Lancashire, Hertfordshire and Derbyshire. Indeed, some 90% of Avon's high technology employment was in that single industry at

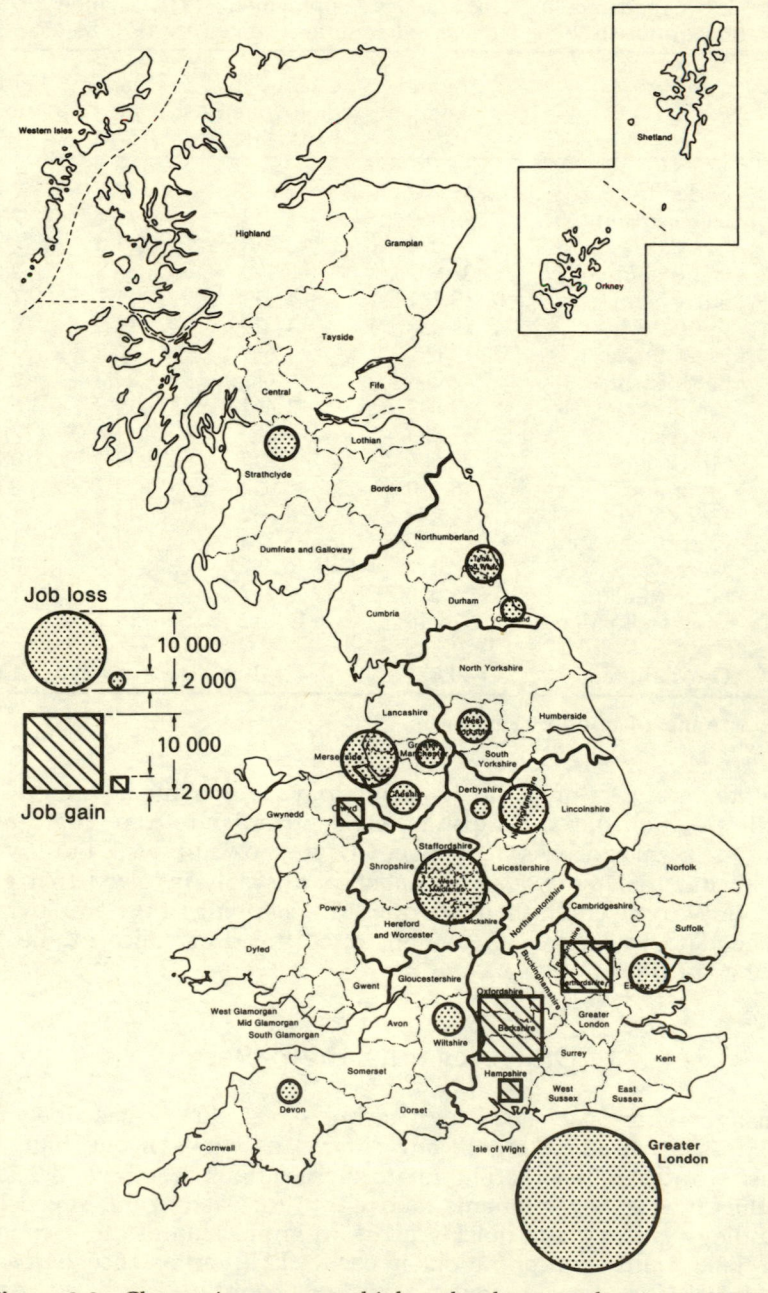

Figure 3.3 Change in aggregate high technology employment, 1975–81 (counties with gain or loss of 2000 or more jobs).
Source: Annual Census of Employment.

Table 3.9 Location quotients for counties for individual 'high technology' sectors.

Location quotient	MLH 272 pharmaceutical chemicals & preparations	MLH 363 telegraph & telephone apparatus	MLH 364 radio & electronic components	MLH 365 broadcast receiving & sound equipment	MLH 366 electronic computers	MLH 367 radio, radar & electronic capital goods	MLH 383 aerospace equipment manufacturing
8.0+				Mid Glamorgan Essex			
7.5–8.0							
7.0–7.5		Fife region					
6.5–7.0							Avon
6.0–6.5							
5.5–6.0					Berkshire Hertfordshire		Isle of Wight Somerset
5.0–5.5	Nottinghamshire West Sussex						Hertfordshire Derbyshire Lancashire
4.5–5.0	Cheshire	Cleveland Nottinghamshire West Midlands					Clwyd
4.0–4.5						West Sussex Fife region	
3.5–4.0	Hertfordshire	Merseyside	Fife region		Hampshire	Berkshire Lothians region Hertfordshire	
3.0–3.5			Essex	Hampshire		Essex Dorset	
2.5–3.0	Kent		Bedfordshire Mid Glamorgan Wiltshire West Sussex		Staffordshire	Surrey	Gloucestershire
2.0–2.5	Merseyside					Kent Hampshire	Dorset Surrey

Note: Only location quotients above 2.0. *Base* = UK.

1981; in Derbyshire and Lancashire the figures were 88% and 77% respectively. It may therefore be more accurate to call these pre-dominantly aerospace centres – or, given that 80% of the British aerospace industry's output is for defence, defence-related high tech centres.

With the other constituent high technology sectors, similar distinct spatial patterns emerge. Pharmaceuticals dominates high tech employment in Nottinghamshire, West Sussex and Cheshire, although in absolute terms it is concentrated in London and Hertfordshire also. Conversely, telegraph and telephone apparatus (MLH 363) jobs are concentrated in the West Midlands and London, but with high relative concentration in Fife, Cleveland and Nottinghamshire. The electronics industries remain concentrated in the South-East, with employment in electronic computers (MLH 366) focused upon London (16% of the national total) and Hertfordshire, Berkshire and Hampshire (with between 10% and 14% each). Surprisingly, some of the popularly assumed high technology areas have very little employment in computers, the archetypal high technology activity: Avon, Cambridge-shire and Wiltshire, for example. In contrast, Staffordshire, of which we hear nothing in all the high technology 'hype', had four times Avon's employment in computers in 1981 and ten times Wiltshire's.

London and the circle of home counties (Essex, Hertfordshire, Hampshire, Berkshire and Kent) contained most of the radio, radar and electronic capital goods (MLH 367) jobs in 1981, although the Lothian region, around Edinburgh, and two further southeastern counties (West Sussex and Surrey) also had large numbers of employees. These southern counties contained over 40% of the national jobs in the industry (60% if London is included). Fife has a high location quotient, although the absolute number of jobs is modest. Radio and electronic components (MLH 364) employment was spread among the main metropolitan counties and the home counties of Hampshire and Essex.

As noted previously, employment change – particularly over recent years – may be a better measure of high technology activity than absolute levels. We have seen that nationally most high technology sectors have been losing jobs, with the exceptions of computers and electronic capital goods. Hence, few counties experienced significant high tech employment gains over the 1975–81 period, and many lost jobs. Figures 3.4–3.10 show major job gains and losses (over 1000) in each of the seven short-listed high technology sectors, over the period 1975–81, by counties. When compared to the aggregate picture in Figure 3.3, they demonstrate very clearly the differing geographies of each sector.

Avon lost 5% of its aerospace jobs and Nottinghamshire lost 30% of its pharmaceuticals jobs over the period. The telegraph and telephone industry experienced a massive 33% national decline, but this was not spread evenly, with the West Midlands doing relatively well (losing

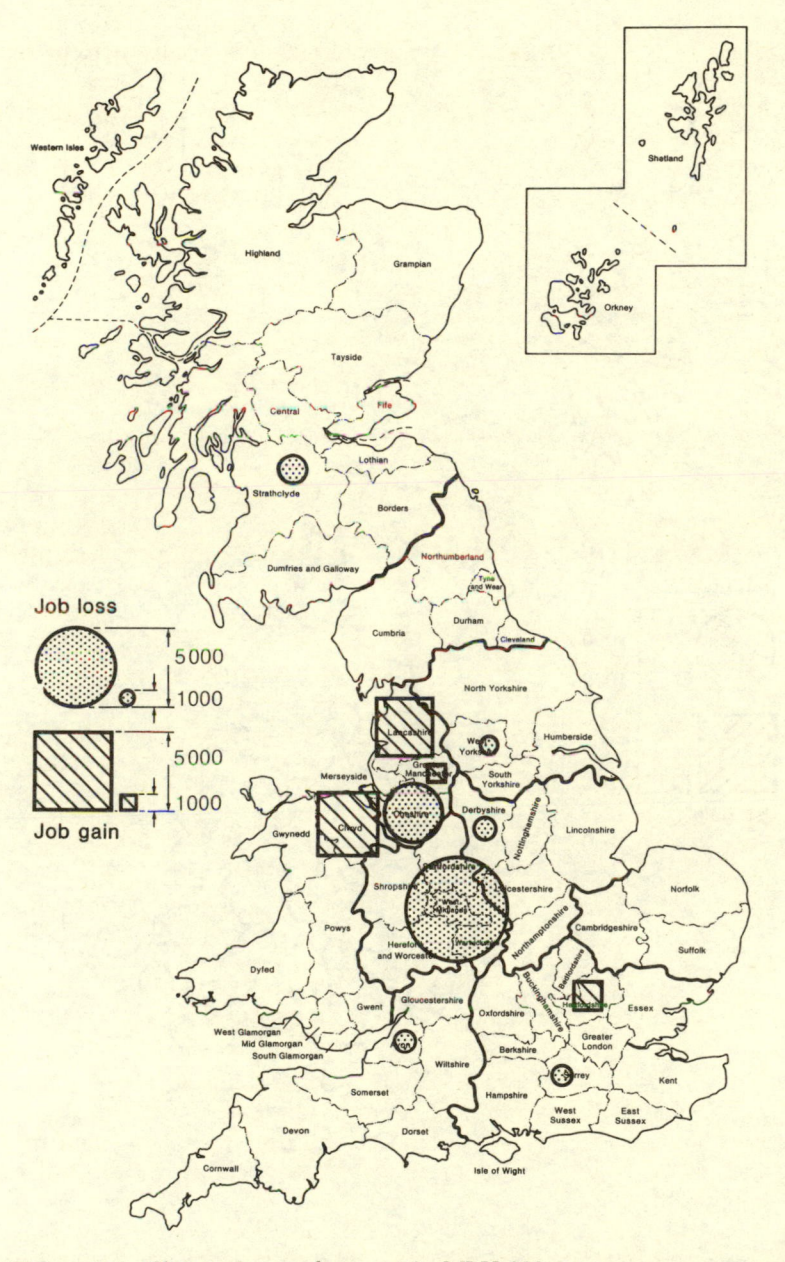

Figure 3.4 Change in employment in MLH 383, aerospace equipment manufacturing, 1975–81 (counties with gain or loss of 1000 or more jobs). *Source*: Annual Census of Employment.

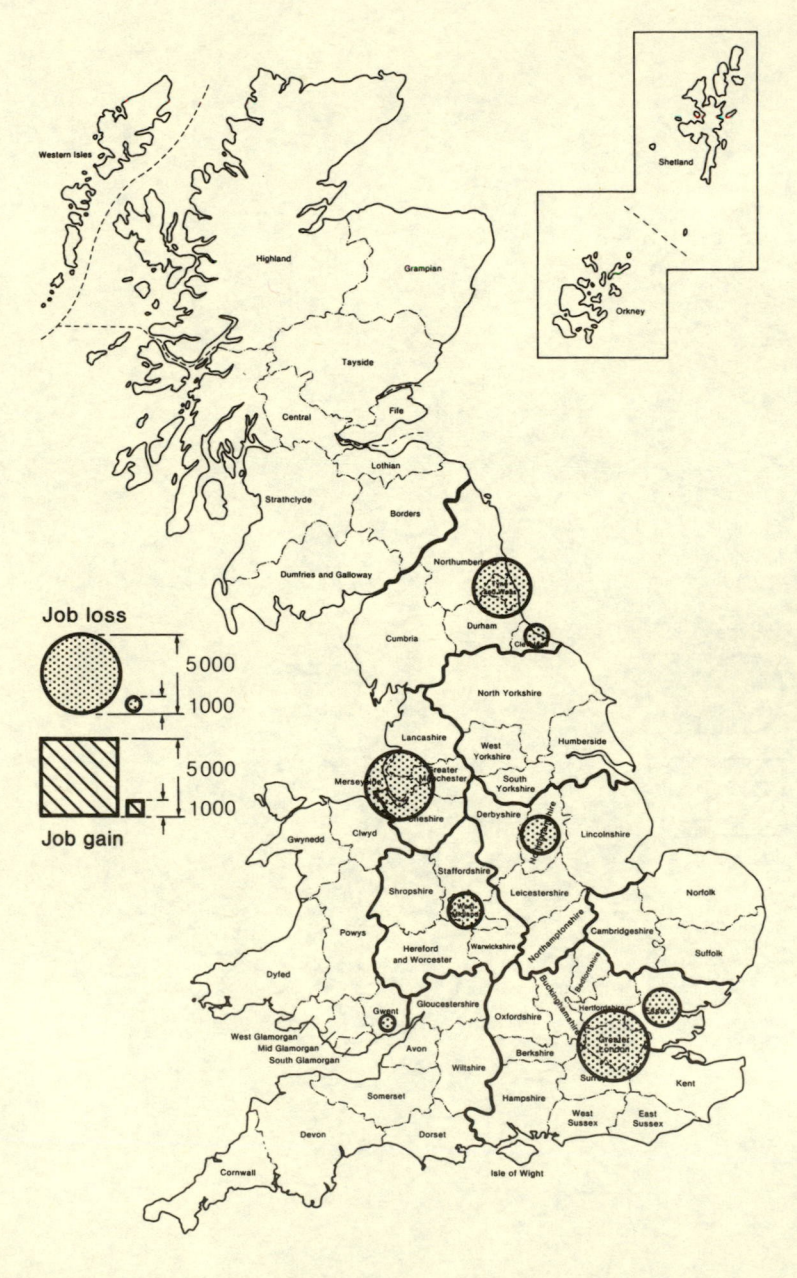

Figure 3.5 Change in employment in MLH 363, telegraph and telephone apparatus, 1975–81 (counties with gain or loss of 1000 or more jobs). *Source*: Annual Census of Employment.

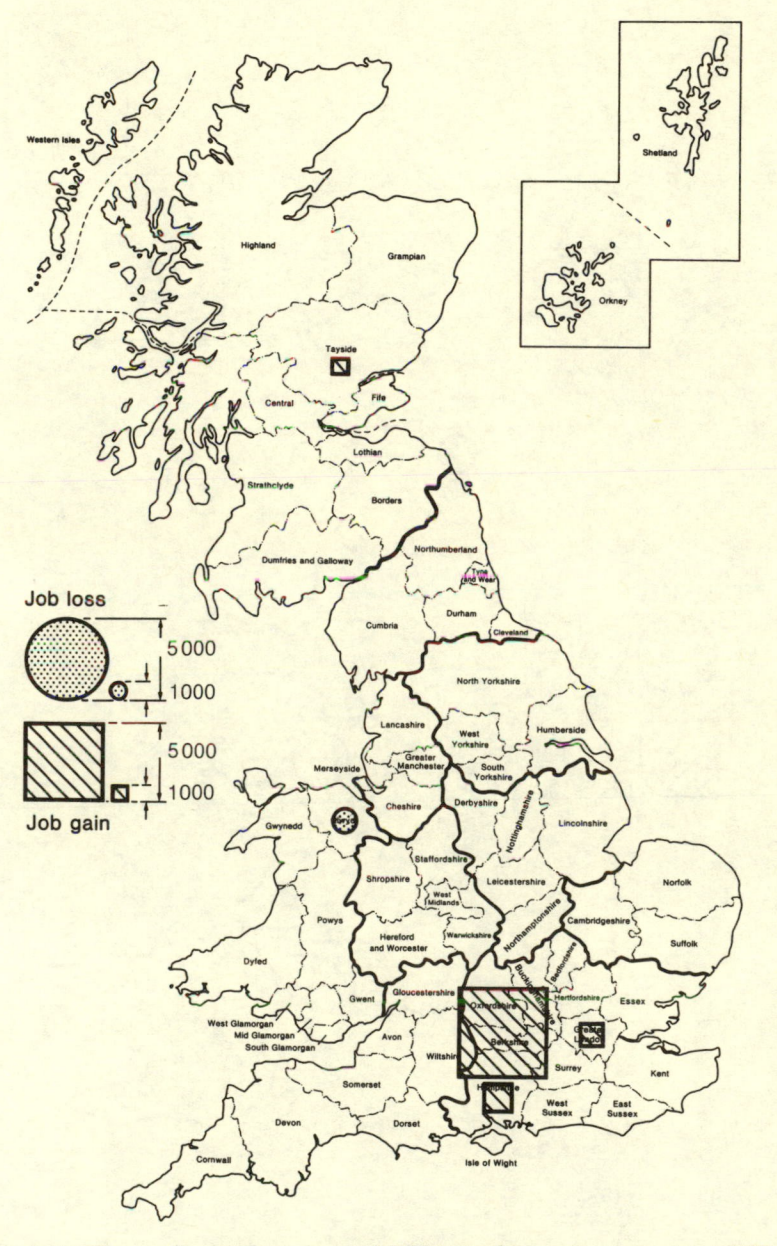

Figure 3.6 Change in employment in MLH 366, electronic computers, 1975–81 (counties with gain or loss of 1000 or more jobs).
Source: Annual Census of Employment.

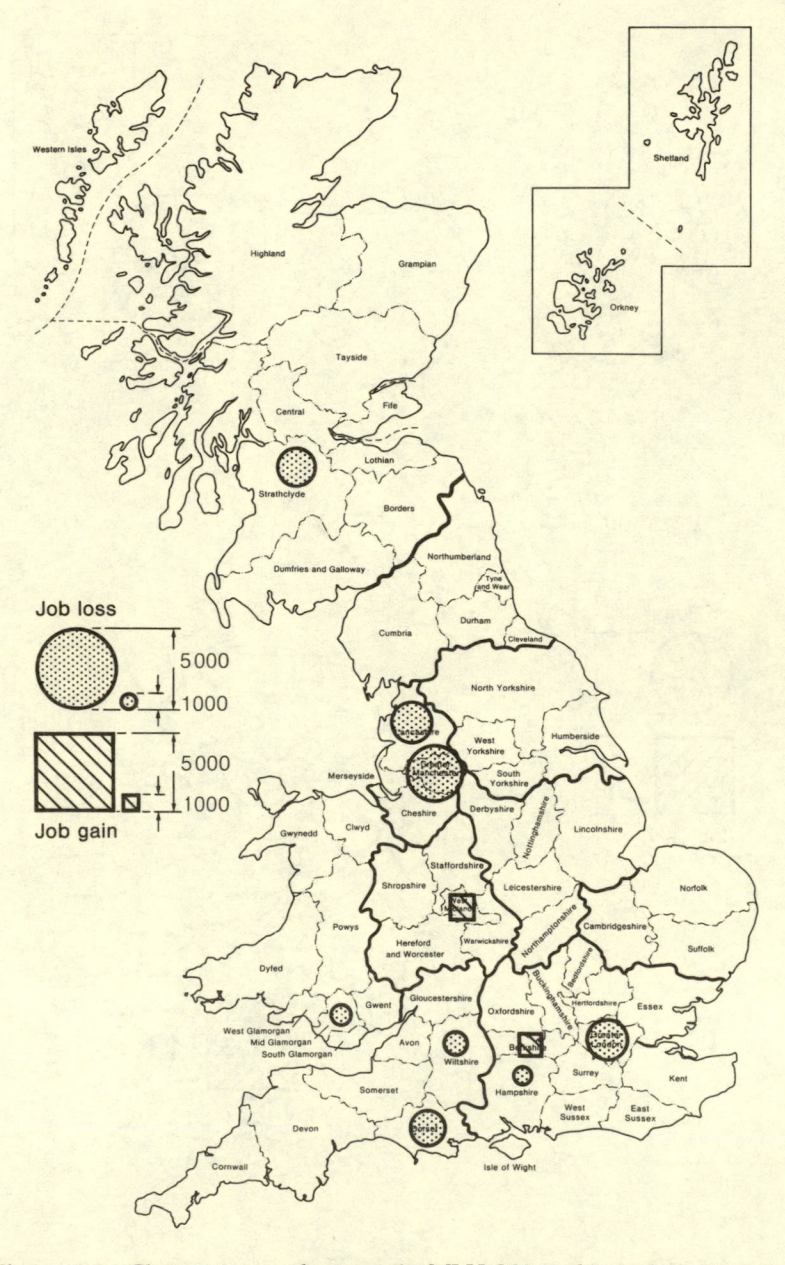

Figure 3.7 Change in employment in MLH 364, radio and electronic components, 1975–81 (counties with gain or loss of 1000 or more jobs). *Source*: Annual Census of Employment.

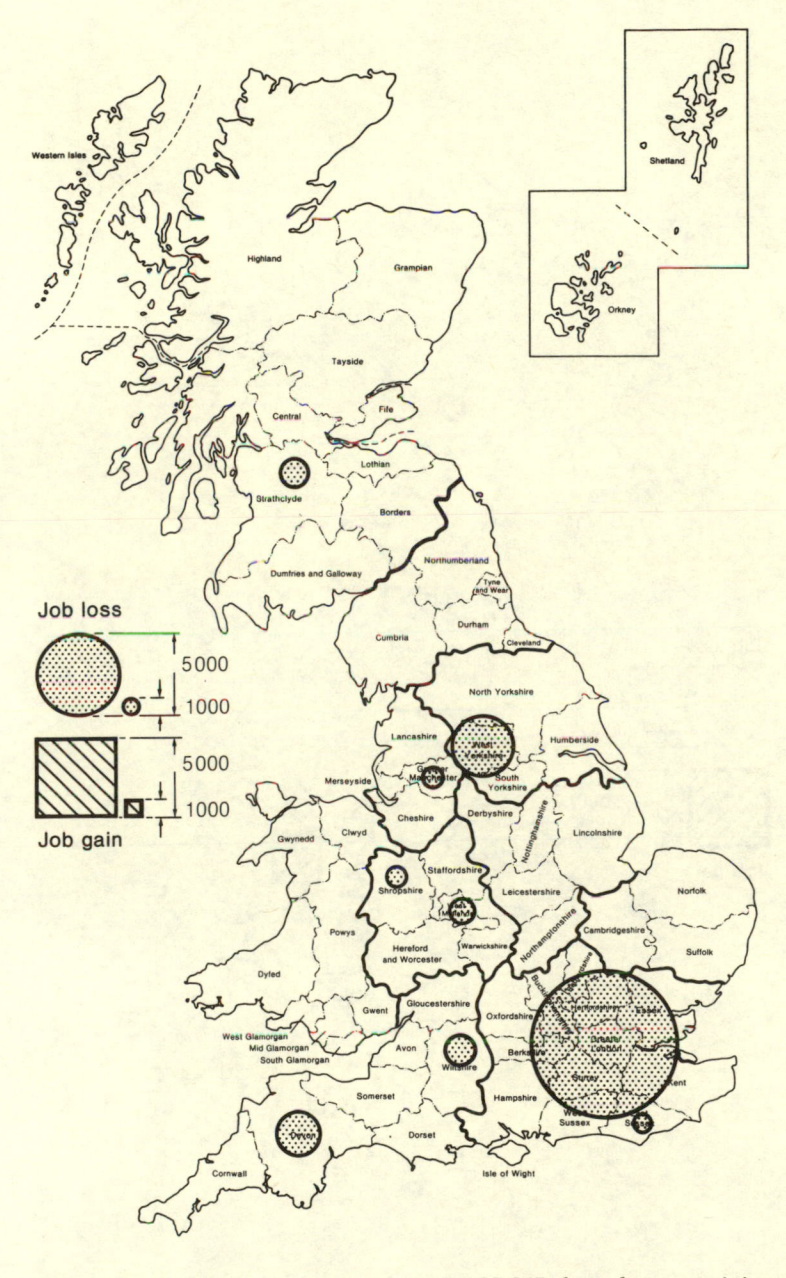

Figure 3.8 Change in employment in MLH 365, broadcast receiving and sound equipment, 1975–81 (counties with gain or loss of 1000 or more jobs). *Source*: Annual Census of Employment.

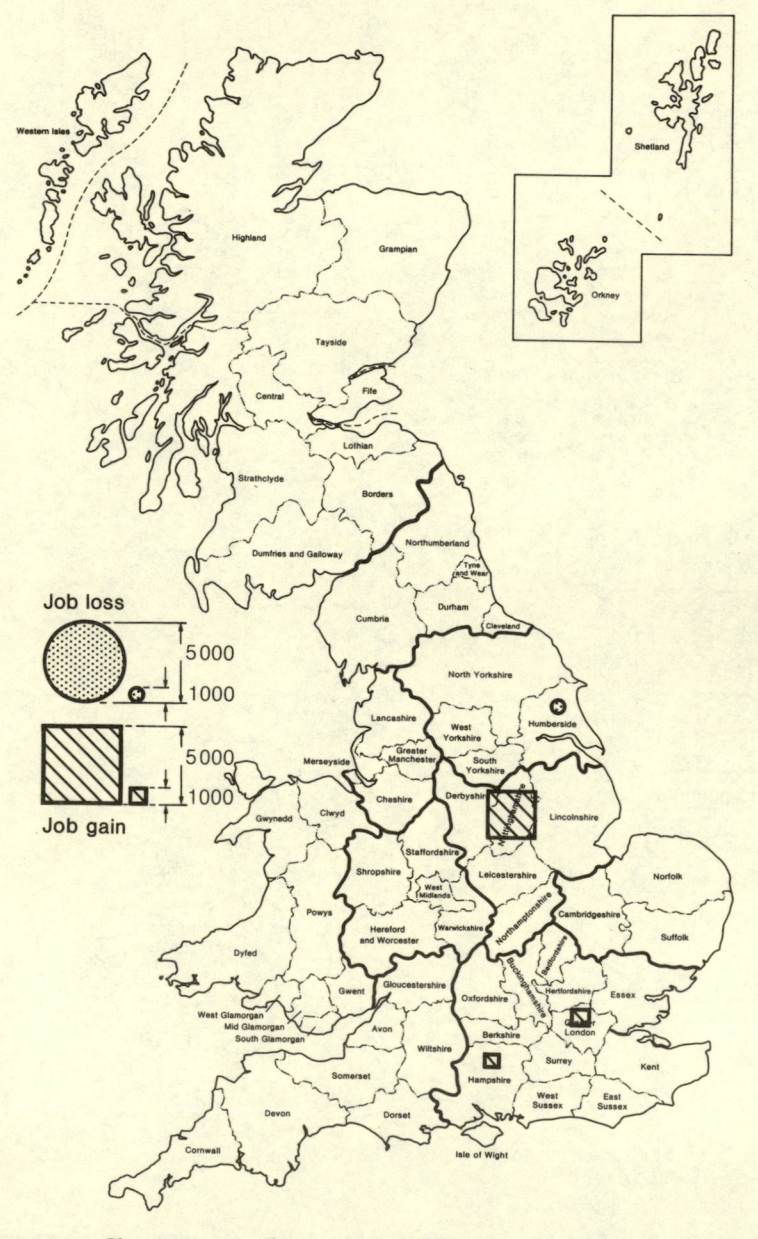

Figure 3.9 Change in employment in MLH 272, pharmaceuticals, chemicals and preparations, 1975–81 (counties with gain or loss of 1000 or more jobs). *Source*: Annual Census of Employment.

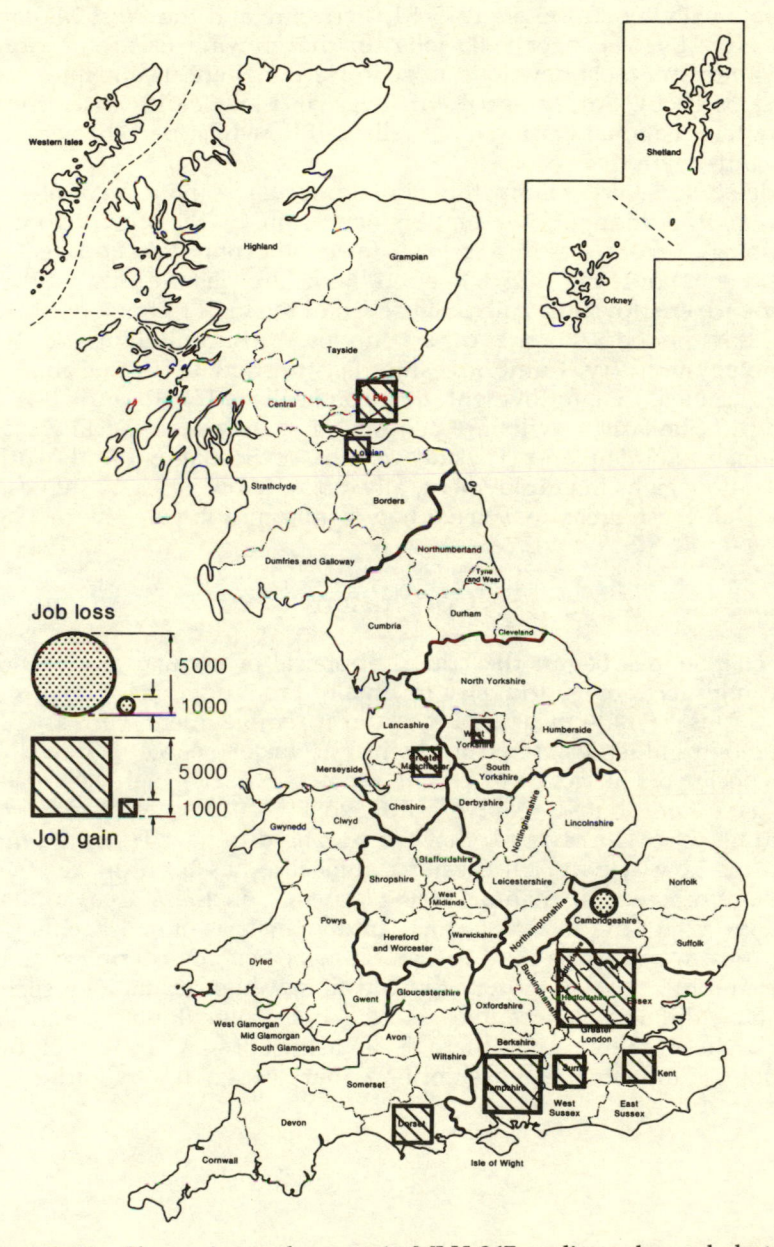

Figure 3.10 Change in employment in MLH 367, radio, radar and electronic capital goods, 1975–81 (counties with gain or loss of 1000 or more jobs).
Source: Annual Census of Employment.

only 14%), compared to London's fall of 29% and Tyne and Wear's of nearly 4000 jobs. While radio and electronic components employment fell nationally by 20% over 1975–81, Berkshire and the West Midlands each grew by well over 1000 jobs. In the growing electronic capital goods industry most new jobs arose in Hertfordshire, Hampshire (over 3500 jobs each), Surrey and Kent (over 2000 jobs each). In electronic computers, national employment fell slightly, whilst Berkshire made a substantial gain.

Indeed, Berkshire, Hampshire and Hertfordshire alone accounted for a large proportion of the employment gain in the only two high technology sectors to grow nationally (electronic computers and electrical components) over the 1975–83 period. Berkshire was the only county to experience employment gain in all seven of the high technology sectors over this period. Of the other supposed concentrations of high technology industry, Cambridgeshire fails to show up in this analysis, with declines in employment over both the 1971–81 and 1975–81 periods. Similarly, Wiltshire shows a substantial decline. The Grampian, Lothian and Tayside regions of Scotland, on the other hand, show gains in employment, albeit from a low base. As suggested earlier, all these areas may have had significant job gains since 1981.

Conclusion

This chapter has begun the critical appraisal of popular assumptions about high technology industry in Britain. From it we may draw three conclusions. First – measured in direct employment terms – high technology industries, far from being the major source of new jobs, have declined nationally in recent years. Secondly, high technology jobs have contributed relatively little to the success of those areas, including the 'M4 Corridor', that are supposed to have benefited most from this new activity. Of greater significance has been the relatively good performance of manufacturing generally in these areas and the rapid growth of the service sector. Thirdly, analysis of high technology industries in aggregate may be misleading when considering patterns of geographical concentration; the constituent high technology sectors each have different geographies. All this casts some doubt on a further popular assumption, that there is a distinct 'M4 Corridor' of high technology industry: an assumption that we go on to address in Chapter 4.

4

Corridor, triangle or crescent?

The obvious question now arises: is there such a thing as a 'Western Corridor' or 'M4 Corridor' at all? Or is it a media and estate-agency myth? In Chapter 3 we looked at various ways of measuring concentrations of high technology industry: absolute concentration, relative concentration, rate of recent growth. We concluded that growth was the most useful measure. On that basis, the statistics fortify the conclusions of Gibbs (1983), Ellin and Gillespie (1983) and EITB (1984): that there is no identifiable, continuous corridor of high technology industry along the M4.

In absolute employment terms, Greater London, at the Corridor's eastern end, with 91 000 workers in 1981, still dominated the pattern despite massive losses in the preceding years. In relative concentration terms, Avon with a location quotient of 2.32 and Somerset with 2.11 scored more strongly than Berkshire's 2.08 or Hampshire's 2.00. But the pattern of 1975–81 growth is dominated by Berkshire with 7600 additional jobs, Hampshire with 2200 and Surrey with 1800 – or, outside the Corridor altogether, by Hertfordshire's 5900 additional jobs.

On this criterion the core of high technology activity, and particularly electronics, in Britain is at the London end of the 'M4 Corridor', in a belt running from Hertfordshire to the north-west of London, through Berkshire and into Hampshire and Surrey. It seems more appropriate, if labels are necessary, to refer to the 'Western Crecent': this pattern of high technology industry is not so much a 'corridor' as an arc around west London. Areas to the west of this crescent, particularly Avon and Wiltshire, popularly believed to feature strongly in the M4 phenomenon, show no particular dynamism; even farther west, South Wales generally shows losses in employment in our high technology sectors, rather than gains. At the county level, then, at least down to 1981, the 'M4 Corridor' is far from a homogeneous entity.

This pattern, interestingly, is also reflected in the geography of rental levels for high technology premises. Figure 4.1, taken from a report by the property agents Fuller Peiser (1985), shows a peak of rentals in the east Berkshire area, centred on Heathrow, Slough and Bracknell. The M4 Corridor effect does show up, but no more so than other areas served by radial motorways from London.

Analysis of Annual Census of Employment data for geographical areas finer than the county level is necessarily limited because of confidentiality restrictions. Thus, for example, we cannot look at each of our individual high tech sectors for any geographical area smaller than counties. However, we can look at aggregate high technology employment changes for functional urban regions (FURs), which are smaller than counties. Figure 4.2 maps changes in high technology employment over the 1975–81 period for FURs in the South-East and South-West of England and South Wales. This confirms the general pattern of change already revealed by the county-level analysis. However, the centres of employment gain and loss can be identified more precisely. Hemel Hempstead, Bracknell, Reading and Portsmouth are the major centres of growth. The major centres of decline are London and, perhaps surprisingly, Harlow, Cambridge, Gosport and Swindon. In South Wales, Swansea has a small but significant gain in an area that otherwise shows general high technology decline.

One major problem with all of this subnational analysis is that it stops at 1981. This is because no official employment data are available at present beyond that date. It is unfortunate because clearly so much activity in the high technology sector has taken place since 1981. This is particularly the case in Swindon and Cambridge, which have only been popularly recognized as high technology centres since that date. Surveys of new firms undertaken by Thamesdown Borough Council, the local authority responsible for Swindon, do not distinguish high technology firms from others. However, recent years have seen a considerable growth of employers in the town – many of whom, we can suppose, will be in our high technology sectors. Likewise, a reading of *The Cambridge phenomenon* (Segal Quince 1985), a study of high technology growth in Cambridge, suggests that much of this is very recent. Figure 4.3, taken from the Segal Quince report, shows that many of the high tech firms in Cambridge have been established very recently. We must await 1984 ACE data before we can update the 1971–81 analysis presented here.

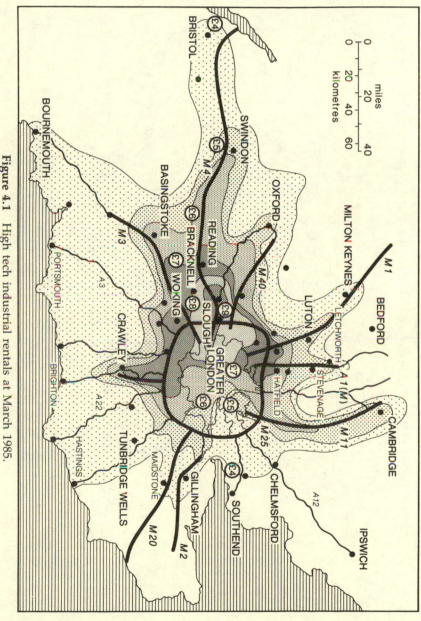

Figure 4.1 High tech industrial rentals at March 1985.

Source: Fuller Peiser (1985).

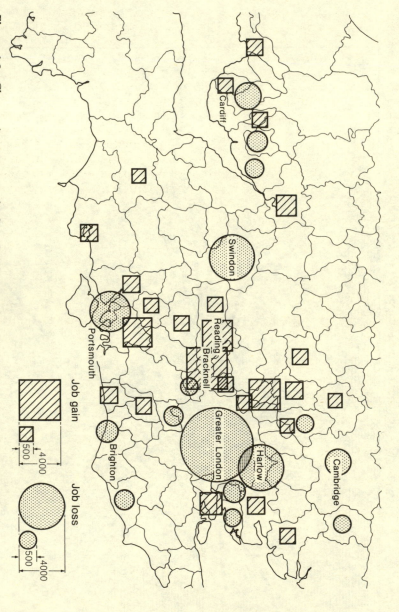

Figure 4.2 Change in aggregate high technology employment for functional urban regions, 1975–81.

Source: Annual Census of Employment.

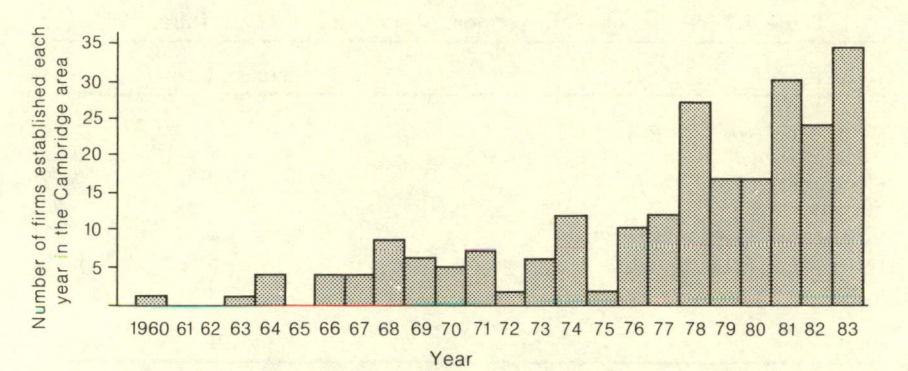

Figure 4.3 Date of establishment of high tech firms in Cambridge.
Source: Segal Quince (1985).

County surveys of
high technology industry

The only available post-1981 employment information, either for counties or for finer geographical areas, comes from surveys undertaken by local authorities. Fortunately, some of the authorities in the M4 area have produced reports on such surveys. Hampshire County Council (1984) has undertaken a survey of specifically 'high technology' firms. Using a broader definition of such firms than has been adopted here (it includes certain service sectors), the survey identifies more than 44 000 jobs in the county (10% of total jobs in the county) in 1984, with 29 500 in south Hampshire and 12 500 in north-east Hampshire. It is in this latter area, immediately south of Berkshire, that the greatest rate of increase has taken place. Just over half of the total high technology employment was found to be in the electronics sectors (Table 4.1). The occupational structure of high technology firms, as shown in Table 4.2, demonstrates the very high proportion of total jobs taken by those in managerial/professional and technical occupations; 92% of the jobs in these categories are taken by males.

The most comprehensive of these surveys is that carried out by Berkshire County Council, with Newbury and Bracknell District Councils, together with some help from the University of Reading, early in 1984. With the exceptions of some small sectors, this attempted a 100% survey of Berkshire employers. The results are very important for our investigation because – as already seen – Berkshire does emerge consistently in our earlier analysis as a major centre of this activity in Britain.

Originally, the Berkshire survey was going to distinguish all firms

Table 4.1   Types of high technology industry in Hampshire, 1984.

Industry	No. of firms	Jobs
instrument engineering	39	5 884
other engineering	17	3 028
electronics	93	22 779
aerospace	10	3 231
business/professional services	23	1 412
research & development	5	6 458
other	4	1 229
total	191	44 021

Source: Hampshire County Council 1984.

Table 4.2   Occupations by sex in high technology industries in Hampshire, 1984.

Occupation	Males	Females	Total	% female
managerial/professional/technical	13 659	1 216	14 875	8
other non-manual	6 901	5 337	12 238	44
skilled manual	6 417	437	6 854	6
semi/unskilled manual	5 311	5 419	10 730	50
total	32 288	12 409	44 697	28

Source: Hampshire County Council 1984.

describing themselves as high tech. Perhaps because of problems of definition, however, the published report (Royal County of Berkshire 1985) distinguishes only electronics companies, which in any event dominate the county's high technology activity. Table 4.3, taken from the report, lists such electronics employment by district and by a three-way classification of firm type. The manufacturing group includes firms engaged largely in production, but possibly involving other functions. The wholesale firms were those in warehouse-type premises undertaking wholesale distribution, plus a mixture of servicing, R&D, administration and software production. The office firms were in office premises, usually engaged in software production, sales, marketing, or pure R&D. Most of the employment in these latter two firm types would not fall within our definition of high technology manufacturing.

The overall estimate of 30 000 high technology (specifically, electronics) jobs in the county at April 1984 compares with a figure of approximately 22 000 from the County Council survey at the end of

Table 4.3 High technology employment in Berkshire, 1984.

	Firms	Manufacturing[a]	Wholesale[b]	Office[c]	Total employment[d]	% of total	Estimate of total	Average size of firm
Newbury	41	981	84	97	1 162	5.8	2 000	28
Reading	77	1 871	1 383	695	3 949	12.3	6 000	51
Wokingham	51	1 057	464	286	1 807	14.6	4 000	35
Bracknell	34	3 408	329	3 187	6 924	30.6	9 000	203
Windsor/Maidenhead	65	193	489	1 164	1 846	8.1	3 000	28
Slough	59	964	1 500	966	3 430	12.9	6 000	58
Berkshire	327	8 474	4 249	6 395	19 118	14.0	30 000	58

Source: Royal County of Berkshire 1985.

Notes: [a] Manufacturing.

[b] Non-manufacturing operating in industrial/warehouse premises, usually engaged in a mixture of wholesale distribution, servicing, some R&D, administration, software production.

[c] Non-manufacturing operating in office premises, usually engaged in software production or sales, and marketing of electronic products or pure R&D.

[d] Of survey *private* sector employment.

1981 (Royal County of Berkshire 1982). Thus, in a period of just over two years, Berkshire's high technology employment increased by approximately 8 000. This compares with an ACE growth of 7 800 for the county over the ten-year period 1971–81. Although the latter figure refers to a more restricted definition of high technology jobs, the comparison demonstrates clearly that post-1981 changes are indeed very significant.

Bracknell, which houses major electronics companies such as ICL, Ferranti, British Aerospace, Racal and Honeywell, has the lion's share of Berkshire's high technology jobs, with the much vaunted Newbury District having the lowest level amongst the five districts (Table 4.3). Interestingly, the well-established industrial areas of Bracknell and Slough have firms which are larger, on average, than those in the other districts. These larger firms, particularly in Bracknell, seem to be functioning as offices rather than as manufacturing plants. Manufacturing features most strongly, as a proportion of total high technology employment, in the more recently industrialized areas of Newbury and Wokingham to the west of the county. In these districts, manufacturing is in small establishments, often producing customized electronic components. Bracknell has by far the largest proportion (30.6%) of its surveyed employment in high technology industries, with Newbury (5.8%) a long way below the county average (14%). If we account for public sector employment, then high technology, using the broad definition adopted by Berkshire County Council (Royal County of Berkshire 1985), accounts for approximately 10% of county employment; this compares with a national figure (in 1982) of 3.8%.

Some 40% of the surveyed high technology companies operate on single sites, with the figure as high as 63% for manufacturing firms. As Table 4.4 shows, 23% of high technology firms are foreign-owned, compared to only 4% for all firms in the survey. Foreign ownership is much higher amongst the warehousing and office activities than in manufacturing. This highlights the distribution functions of many foreign firms.

Table 4.5 shows that a large proportion of single-site firms (53%) are

Table 4.4 High technology industry in Berkshire: function of site by activity (percentage of firms).

	Manufacturing	Wholesale	Office	Overall	All firms
no other site	63	29	32	40	50
HQ site	13	19	22	18	11
branch, HQ in UK	15	20	20	19	34
branch, HQ abroad	9	32	26	23	4

Source: Royal County of Berkshire 1985.
Note: Not including Bracknell and Newbury.

small, having nine or fewer employees. Surprisingly, 27% of the smallest (1–9 employees) firms are foreign-owned. Again, these are likely to be distribution centres or offices for foreign manufacturers.

The survey confirms that much of Berkshire's high tech activity is recently established. Figure 4.4 shows that a much higher proportion of electronics companies have been established in the last five years than is the case with all firms. Table 4.6 provides a more detailed assessment of age according to the function of sites. The foreign-owned element is consistent over the age range, varying between 19% and 27%, suggesting no sudden influx of foreign firms over the last 25 years.

One finding supports the view that the nature of high technology industry in Berkshire has changed in recent years. Figure 4.5 distinguishes older firms, established for more than five years, from newer ones. Only 20% of newer electronics firms are involved primarily in manufacturing, compared to 48% for older firms. The converse is that a high proportion of new electronics firms are functioning mainly as offices (49%), compared to older firms (27%). This supports the view that as Berkshire's high tech role has grown, it

Table 4.5 High technology industry in Berkshire: function of site by number of employees (percentage of firms).

	Numbers of employees				
	1–9	6–10	25–99	100+	Overall
no other site	53	44	31	11	40
HQ site	7	18	26	36	18
branch, HQ in UK	13	17	20	39	19
branch, HQ abroad	27	21	23	14	23

Source: Royal County of Berkshire 1985.

Table 4.6 High technology industry in Berkshire: function of site by age of firm (percentage of firms).

	Age of firm (years)			
	1–5	6–10	11–24	25+
no other site	43	45	32	*
HQ site	18	21	15	*
branch, HQ in UK	12	14	28	*
branch, HQ abroad	27	19	24	*

Source: Royal County of Berkshire 1985.
*Sample size too small.

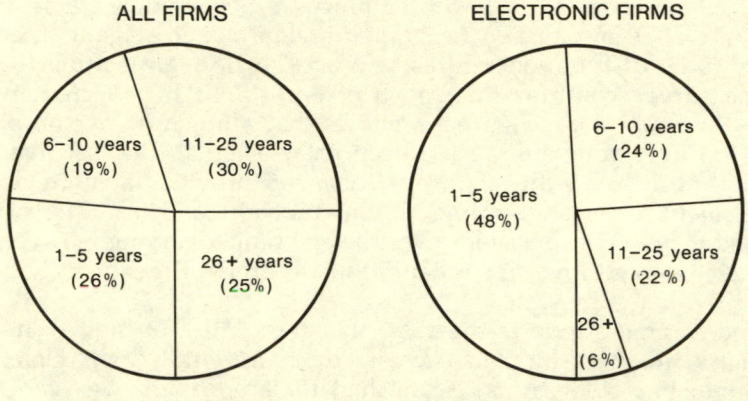

Figure 4.4   Age of electronics and all firms in Berkshire.
Source: Royal County of Berkshire (1985).

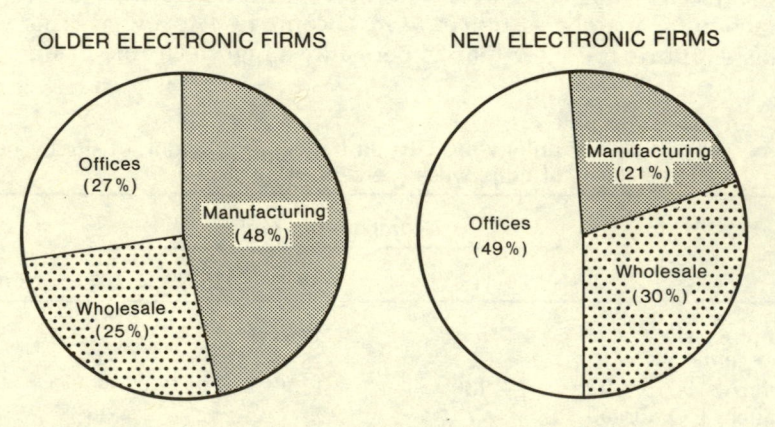

Figure 4.5   Age of electronics firms in Berkshire by function.
Source: Royal County of Berkshire (1985).

has increasingly taken the form of administration, R&D and marketing, rather than manufacturing.

This changing role is reflected in the occupational structure of the electronics sector. It is now conventional wisdom that high technology sectors will have a high proportion of their workforce in professional and technical occupations. However, the Berkshire survey shows that here this bias is very marked. Figure 4.6 shows that professional/ technical workers account for 42% of total electronics jobs (rising to 58% for the growing office function), compared to 20% for all firms. Only 12% of electronics firms jobs are taken by non-skilled manual workers, as against 29% for all firms responding to the survey. Even in electronics manufacturing firms the figure is only 19%.

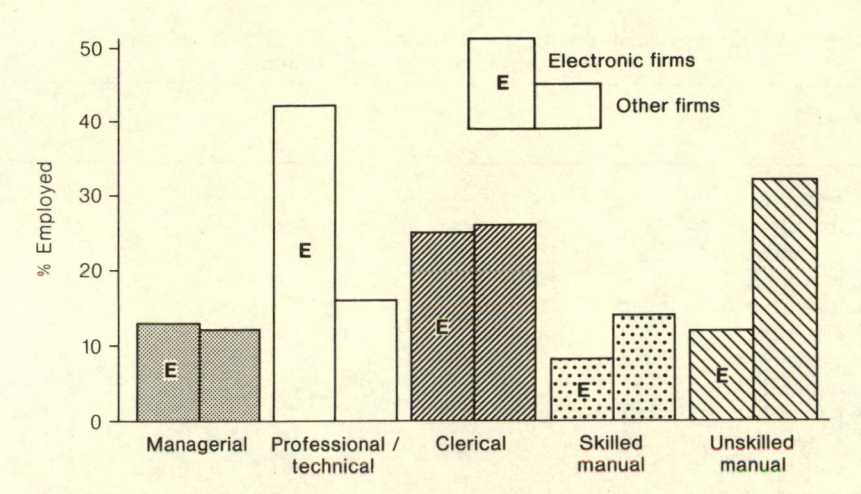

Figure 4.6 Occupation structure of high tech firms and all firms in Berkshire. *Source*: Royal County of Berkshire (1985).

The multiplier effects of high technology industry

Nationally, we have seen in Chapter 3 that the role of high tech as employment creator has been greatly exaggerated. The same goes even for Berkshire. Although in recent years the county has had the highest rate of high technology growth in Britain, the direct job creation effect has been relatively small. Table 4.7 shows 1971–81 employment change for selected industrial sectors. Although manufacturing declined in aggregate, a number of individual sectors – notably electrical engineering – showed quite strong growth in comparison with the country as a whole or with other counties. That said, the major employment growth came from the service sectors. These together contributed a net gain of 49 000 jobs over the decade, compared to a growth in our high technology sector of 7 500.

Nevertheless, high technology industry may create particularly large secondary, indirect employment effects – larger, perhaps, than other sectors do. Berkshire's success in the service sectors, for instance, might be just such an indirect result of growth in the high technology sectors. We cannot answer this question rigorously. But we can speculate as to the likely links between high technology and other sectors, and offer what little evidence is available.

High technology sectors are likely to have all of the multiplier effects on a local area that we would expect from any manufacturing sector. They will consume goods and services from other sectors of the local economy, and their employees will spend part of their income on

Table 4.7 Employment change in Berkshire, 1971–81, for selected sectors (1968 SIC, employees in employment).

SIC	Change 1971–81	Percentage change
1 agriculture	−1 630	−32.6
3 food	−2 103	−21.6
5 chemicals	−978	−11.1
7 mechanical	−4 505	−21.5
8 instrument engineering	127	4.0
9 electrical	7 283	60.2
11 vehicles	−230	−4.1
12 metal goods	−582	−7.9
18 paper, etc., printing	−827	−8.4
19 other manufacturing	336	9.1
20 construction	−731	−5.0
21 utilities	540	12.8
22 transport	−1 858	−12.4
23 distribution	12 391	34.4
24 banking	11 672	104.1
26 miscellaneous services	12 137	45.8
28 private, professional & scientific services	7 536	86.1
29 public services	9 734	11.7

Source: Adapted from SERPLAN 1985.

goods and services produced locally. Beyond this, do we have any reason to assume that multiplier effects will be any larger for high technology sectors? Possibly we do. One feature of high technology industries is that they are relatively capital-intensive, and so employ fewer workers per unit of output than many other sectors. Thus, we might expect high technology companies to create demands for goods and services, and generally generate wealth, out of proportion to their direct employment levels. An obvious feature of high technology companies that will tend to produce stronger than average multiplier effects is the fact that a high proportion of workers in high technology companies are in managerial, professional and technical occupations, implying high salaries and spending power. In addition, in an area such as Berkshire, where high technology companies have congregated, we might expect that a good many components required by manufacturers can be supplied locally, thus reducing the need for imports and boosting the local economy. Indeed, a feature of the electronics industry in the area is the strong links between contractors and subcontractors.

A less tangible feature of high technology companies that may produce higher than average multiplier effects is the image or cachet that they seem to give to an area. There is ample evidence for this: local authorities have desperately tried to attract mobile high technology companies by such devices as science parks; property agents have obliged by persuading investors to put money into high technology industrial developments and prospective tenants to occupy 'high tech' buildings. The process may be self-fulfilling, as decision makers – businessmen, investors and government – acquire an increasingly favourable view of the area. Thus, the area becomes attractive not only to further high technology companies, but also to other companies, such as producer services decentralizing from London.

Such speculations must remain strictly that, unless and until someone attempts some form of local input–output analysis. All we can say is that even in the most high tech part of the Corridor, its direct contribution to the growth of employment has been relatively modest. Further, even within the high tech sector, much of the employment in the core of the Corridor appears to be of an administrative or research character. These facts need to be kept in mind as we go on to look in more detail at the factors that contributed to the establishment of high tech firms here. In Chapter 5, we report on the empirical results of our own follow-up survey of Berkshire high tech firms. In Chapter 6, we begin to develop a tentative theoretical explanation for the high tech phenomenon, which we expand and test in subsequent chapters.

5

Sunrise companies

To discover how Berkshire's firms had originated and grown – and particularly, how and why they had grown in that area – we interviewed a substantial sample during 1984, chosen randomly from all the high tech firms in the county. Our very long interview questionnaire was designed to probe their history, and in particular the critical reasons for their locational choices. In all, we interviewed 44 firms; for one reason or another only 40 produced fully comparable research findings.

There is another qualification that must be borne in mind when interpreting the answers. We conducted the interviews early in our study, before we had any fully developed theory of location. That was because we wanted to use the results to develop such a theory: a process we describe in Chapter 6. Because of this fact, and because we wanted to probe the reasons for location in some depth, we deliberately asked open-ended questions and then gave our respondents a great variety of possible factors. Sometimes these worked at more than one level: thus, on the significance of transport factors, we first asked the open-ended question, 'Why did you locate here?', then offered a variety of transport factors (the M4, other motorways and trunk roads, Heathrow Airport) and then asked about their significance (for supplies, for shipping product, for customer access, and so on). In practice, respondents were not always able to give a precise answer; thus they might cite 'communications' rather than a specific feature. And sometimes when we repeated a question to get further elucidation, we found that we did not get consistent responses. In what follows, nevertheless, we have tried to quantify the answers. But this has involved a degree of value judgement on our part in interpreting the replies; and the reader must keep this in mind.

The 40 firms divide very obviously, but also very significantly, into three distinct groups (Table 5.1). *Type 1* embraces 22 firms, still generally quite small, whose site in the area is their only operation. Most though not all are fairly recent in origin. The majority – 15 out of 22 – are in manufacturing, though many combine this with a research function, and virtually all are by definition also sales and administrative operations. *Type 2* is a much smaller group of five British firms with more than one location. They are on average distinctly longer-established in the area, and they are much bigger – some of them very big indeed. They are about evenly divided between manufacturing and sales functions. *Type 3* consists of 13 subsidiaries of foreign multi-nationals – nine American, two Swedish, one Swiss, one German. They are on average intermediate in their date of establishment here between the small single-establishment and the big multi-establishment British firms. They are predominantly in sales, or in office-based administration, rather than in manufacturing.

Type 1: the single-location firms

The single-location firms tend to display a very common pattern. All but five of them are fairly young, having been started since 1970. Most are still run by their founders, and remain very small; only three have more than 30 employees. All but three were founded through breakaways by employees of existing firms, predominantly large high tech firms (9 out of 19 cases); the remaining three were founded by freelance contractors to existing firms. In all but four cases, they were located where these founders happened to live; if they have subsequently moved, it has been locally, in search of more space or more suitable premises. Their directors cite customer access and residential preference as the most important locational considerations, followed by staff availability and suitable premises.

For these firms, it is clear that the real explanation of the location is the siting of the original firm from which the breakaway took place – and with which the new firm would maintain either a competitive relationship, serving the same customers, or in rarer cases a continuing subcontractual relationship. In other words, we are dealing with a fairly classic case of a local industrial complex growing through 'spawning' or spin-off effects, as Rogers and Larson (1984) have described in their genealogical tree of Silicon Valley. Typically, the founder of a new firm would have worked for a large firm in London and found an affordable house in Berkshire, becoming a commuter; or the large firm might have been in Berkshire. Logically, in starting up his own firm he would look for a site within easy travel distance of his home. The process is thus part and parcel of the growth and spread of the West London high tech industrial complex – a

Table 5.1 The three types of firm.

	Type 1 UK-based single-site	Type 2 UK-based multi-site	Type 3 Multi-national multi-site
number of cases	22	5	13
year of foundation here			
range	1947–83	1946–76	1948–82
mean	1972	1963	1967
number of staff here			
range	5–144	20–2631	12–1010
mean	25	601	172
main activity here			
manufacturing	15	2	2
sales & servicing	5	2	7
research & development	1	–	1
office/administration	1	1	3
original (UK) location			
here	9	–	5
Berkshire	9	1	4
home counties	2	2	3
elsewhere in UK	2	2	1
origin of founder			
large electronics firm	9	NA	NA
small electronics firm	5	NA	NA
non-electronics firm	5	NA	NA
freelance/contract worker	3	NA	NA
lived here yes	18	NA	NA
no	4	NA	NA
location factors ranked	1 customer access	1 other motorways	1 Heathrow
	2 residential preferences	2 M4	2 staff available
	3 staff available	3 customer access	3 premises
	4 premises		4 customer access
staff: % managerial/ professional			
range	10–93	20–100	17–83
mean	42	49	49
recruitment: % high tech			
range	5–50	0–50	10–100
mean	36	33	73
business: % defence work			
range	0–90	0–100	0–42
mean	19	31	20

Source: Company interviews.

process to be described in Chapter 6 – reinforced, but only in a small minority of cases, by the need for access to the Government Research Establishments for defence contracts.

Some case studies

Firm 1 makes equipment for image processing, which uses computers to extract information from pictures. Founded as recently as 1981, it is located in Wokingham with 25 employees. Its founder, trained as an engineer at the University of London, developed a unit at a major electronics manufacturer for signal processing and then took over an existing unit at another major, London-based firm in image processing – both within commuting distance of his home in Maidenhead. When this firm was taken over by another giant, who was less interested in this field, he broke away, soon bringing six designers with him. The critical consideration for location was access to GREs at St Albans (particularly important), Portsmouth, Weymouth and Malvern. 'The face of the St Albans contract negotiator dropped when it was suggested the firm might move away' to Cornwall or Wales. Access to Heathrow is useful for meeting clients, but is not essential.

Firm 5 makes electronic instrumentation, particularly for the water industry. Located at Woodley on the east side of Reading, it has only nine staff. It was founded in 1973 by an engineer then working at yet another major manufacturer, who developed his own ideas for instruments and seized the chance of finance from an individual. His main consideration in locating was that he lived in Maidenhead and his daughter was at school in Reading. The location is adequate, though a site more central in the country might be better; labour is hard to find because of housing costs. Heathrow is not very important.

Firm 7 is at Twyford near Maidenhead, with 16 employees. It makes vibration monitoring apparatus for use in conjunction with equipment like power station turbines. It is a spin-off from a firm founded in Wokingham in 1962 by three ex-employees of an American factory, originally making laboratory equipment but then branching into industrial applications, which closed. In 1981 the sales director bought out this latter development, locating in Twyford so as to retain staff.

Firm 8, making specialized electronic and optical instrumentation, is located in Reading with eight employees. It was founded in 1974 by a sales and a design worker in an existing small electronics firm, starting in the garage and the garden shed of their respective homes. After three years of struggle they got a contract from the Royal Signals and Radar Establishment, Malvern. They then developed an interface from their system to a desk-top computer. They now sell applications knowledge rather than just hardware. They located where they lived, but find the area good for specialized labour (electronic assembly) and access to component suppliers.

Firm 10, in Reading with nine employees, makes pressure-sensitive digitizers with applications in both the home and school (transfer of children's drawings to computers) and in industrial design. It was founded in 1982 as a breakaway by an employee of another small electronics firm, who had been educated at the University of Reading and made his home in the town after graduation. No other location was considered, but the site has cost advantages and university graduates provide qualified labour.

Firm 12 was founded in 1975 to make transducers and transducer systems. Located in Wokingham, it has 20 staff. It was started by three workers in two existing big firms, who decided to specialize in selling this equipment and later moved into making it. Wokingham was equidistant from their homes. They did not consider other locations, but find this one good for access to local component suppliers and local customers; Heathrow and the M4 are positive factors. About 20% of work is for GREs, including Harwell and Aldermaston; 'you may not get the contract in the first place if you are not here', and the work helps development of commercial spin-offs.

Firm 20, an electronics subcontractor in Reading with nine workers, was founded in 1976 by an Uxbridge-trained engineer who had worked worldwide but returned to Reading where his mother lived. He originally set up business in a potato shed.

Firm 23, with 29 workers in Reading, was founded in 1975 by a freelance draughtsman to make printed circuit boards, but is expanding into other products. This was his home base, but the location is central for the electronics business with a stable, non-militant staff and access to customers – most of whom are within 30 miles (48 km), including some GREs.

Firm 29, located in Reading with 14 staff, makes specialist software for the motor trade. It was founded in 1973 by a redundant business consultant in London, who moved here in search of lower rents. It had a troubled early history until 1978, when the former partner, who had left for a major motor manufacturing company, recommended the product to its dealers. Reading was suitable for labour and communications, and had good cheap office space.

Firm 31 makes electronic equipment and systems for civil and military uses. It employs 27 people in Slough and was founded in 1972 by two contract engineers from a major electronics firm who lived nearby. It has made one local move.

Firm 39, in Reading with 11 staff, provides education and training in operation of large computer systems. It was started in 1981 by three directors, two of whom had previously worked for a large Maidenhead-based firm in a similar business. They could theoretically locate anywhere, but stayed here because it is convenient for London where much of the business occurs.

Firm 42, which specializes in machine tool retrofitting, has 29 workers at its Slough base. It was founded in 1971 by three workers in a High Wycombe firm which had been taken over by one of the major manufacturers, who proposed to move the work to Dorset. They did not want to move and doubted whether the large firm could succeed in this specialized field. They moved to Slough in 1980 to get more space.

Type 2: the UK-based multi-site companies

This small group is very different. The firms are older-established in the area – three out of the five before 1965 – and tend to be even older-established at their original base. They range from the small to the gigantic. They are equally heterogeneous in other ways. Two of them, including the biggest, are heavily involved in defence applications for which nearness to the GREs is important. With one exception (a firm that was taken over) these firms seldom originated in the local Berkshire area, but have headquarters outside it. In their present Berkshire location they place particular importance on the motorway network for reaching their customers, who may be very influential procurement officers. Beyond this, it is not possible to distil a specific set of reasons for the large firms locating along the M4 Corridor. But certain themes did recur in the interviews: the role of out-migration from London, the attraction of Heathrow, the perception of the area as a 'pleasant place to live' and the availability of skilled labour.

Some case studies

Firm 13 is a software information systems contractor, which provides software solutions to particularly difficult problems. Founded in 1962, it has six branches including the London headquarters, with a total staff of 248. The Reading branch, opened in 1966, serves south and west England; an office at Alderley Edge in Cheshire serves the north. It was located here, first, because a number of workers lived in the area and commuted to London; secondly, because of important contacts with three GREs, AERE Harwell, AWRE Aldermaston and RAE Farnborough, who subcontract difficult state-of the-art problems (such as testing Concorde in the early 1970s). Defence work, however, represents only some 5–10% of business. Communications, including the M4, M3 and M25, and the rail network, are also a significant advantage today. Labour availability and a good environment come in second place; the company finds it difficult to recruit in competition with the bright lights of London.

Firm 16 specializes in an overnight delivery service for the computer trade. Founded in 1967 in Manchester, it maintains its headquarters in Reading (with a staff of 20) but now has some 11 regional units. The founder came to Reading in 1977 because he was already contracting for a major American multinational when they moved here from Manchester; he worked exclusively for them until 1982 when he began to develop a nationwide service. The headquarters could be 'anywhere

in the UK but preferably in the Thames Valley area' because of access
to local customers plus good communications.

Firm 38 specializes in the design and production of thick-film
microcircuits. It originated as long ago as 1946–7 in Fleet, Hampshire;
its founder was the first to develop printed circuits in Britain and
possibly the world. Due to expansion it moved in the early 1970s to
Bracknell. It was taken over in late 1980 by a Billericay-based group
with sites all over the country, and a microcircuit function moved in.
Ironically, the market then collapsed and much of the traditional work
was abandoned with a loss of 150 staff. Today the unit employs 70
people.

Firm 42 is the divisional headquarters of a very large aerospace group
headquartered in Stevenage. The site in Bracknell, opened in 1960,
employs no fewer than 2631 people and provides divisional functions:
sales, production support, technical publications, design and development work with some prototype manufacturing plus testing and
quality control. The immediate reason was the need to bring together
some 5500 employees in defence applications ('cost-plus' work) then
scattered in Sudbury, Brentford and Stonehouse (Gloucestershire).
Bracknell was chosen because of its proximity to the GREs (RAE
Farnborough, ASWE Portsmouth, ACO Slough and Ministry of
Defence HQ London); it was possible to make easy day trips; if the
firm were 100 miles (161 km) away, they would get 'half the visits from
MOD staff'. Personal contacts were critical; if the relevant GREs and
procurement section of MOD moved to Liverpool, much of the firm
could eventually follow. The availability and quality of housing were
another important factor.

Bracknell Development Corporation, ironically, did not welcome the
firm because of fears that the new town would become a one-industry
town, but accepted it under central government pressure. Their fears
proved right after the cancellation of Blue Streak caused a reduction in
workforce from 5500 to 2500. At this point the company played on
Bracknell's unemployment worries to secure the transfer of the
Brentford division. Today, easy availability of the right quality of
labour is important; so is easy contact with customers via the
motorway system (though in 1960 there was no M4), and the nearness
of Heathrow for growing international trade.

Type 3: the multinationals

For the Type 3 firms (as indeed for the Type 2 firms) it was often more
difficult to discover the reasons for the initial location, since this had
happened many years before and it was hard even to find who had
made the decision. Generally these are firms, sometimes very large and
well-known, that took a conscious decision to set up a British (or
European) base for their sales and servicing operations; manufacturing,

where it exists at all, is often restricted in scope, consisting of the assembly of knocked-down subassemblies. Their UK base varies greatly in size; the bigger ones may have more than one unit, and tend to have relocated as they grew. They generally made a conscious search for a site near to London and they place particular emphasis – much more so that the indigenous firms – on access to Heathrow for importing products and parts, and for access to European markets. Since they tend to utilize a high proportion of highly qualified sales and servicing staff, however, they also place particular weight on its availability in this area. Though most have little or no concern with defence applications, one or two of them maintain close relationships with the local GREs.

Those multinationals that do have an R&D and a manufacturing presence in Britain often put the R&D – together with office functions – in the South-East, and production elsewhere. Thus one major American firm has four Berkshire sites covering administration, R&D, sales and servicing. Production, however, is carried out at South Queensferry near Edinburgh and near Bristol. Another has 800 employees at its UK headquarters offices in Reading, including sales, service, administration, field service logistics and computer services design and manufacture; in addition, six other Reading sites employ another 340 people, while 250 are with customer support and sales at Basingstoke; but the UK production plant employs over 500 people at Ayr in Scotland.

Some case studies

Firm 17 is a British subsidiary (employing only 12 people) of a very large and old-established German electrical company, set up here in 1948 to specialize in medical engineering in the fields of radiology, electromedicine, dentistry and electro-acoustics. The British headquarters is at Sunbury on Thames; the Reading office is a regional branch for Oxford and Wessex National Health areas. Communications with the firm's manufacturing unit in Britain, and for goods imported from Germany and Scandinavia, are seen as critical factors.

Firm 18 is an office-manufacturing subsidiary of an American company manufacturing access control (security) systems, established in Britain in 1965. The UK founder began as an agent for the US parent, working out of a very small office in the western part of Greater London. The move to Reading in 1968–9 came because of the need for more space and lower rents; since then the firm has moved and expanded. Heathrow is a critical locational factor for supply of materials and components from the US and export to Europe and the Middle East; availability of suitable-quality staff is also crucial, though they are in short supply here as elsewhere. There are no links with GREs. The unit now employs 100 people.

Firm 19 is the subsidiary of a major American firm which pioneered the development of magnetic video-recording material. Located in Reading, with a staff of 150, it is basically a distribution depot with minimal manufacturing functions. It started here as early as 1959 because the UK, with its well-established TV industry, was the first European market to expand. Apparently the nearness of Heathrow was crucial in location; in any event, the managing director came from Sonning. Today the area is thought convenient to widely scattered customers, but by no means essential; proximity to customers is not very important, nor is supply of labour. Heathrow is still much used for importing goods, and is regarded as the most important locational factor. MOD contracts are significant (15–20% turnover) and Reading proves a convenient location for reaching them.

Firm 21 is the British subsidiary of an old-established and well-known Silicon Valley electronics firm. The site at Winnersh, near Reading, employs 1010 people and is the UK headquarters for the measurement systems and computer sales groups, concentrating solely on sales and administration; three others of the company's divisions have sales centres elsewhere due to lack of space on the current site. There is also a customer support centre nearby, and branches at Bracknell, Crowthorne and Pinewood. Most manufacturing for the entire world market is still concentrated at the firm's California base. A developing complex at Bristol will soon manufacture peripherals and disc drives as well as housing a high-level 'Think Tank'; a customer response centre will be located at Pinewood.

The company's first British unit, in 1962, was at Bedford. The Winnersh operation began in 1973–4 and the nearby service centre in 1983. Since the company is traditionally R&D-based, it was attracted here by high-quality labour and access to universities as well as the generally pleasant environment. The Berkshire sales headquarters was, however, located for nearness to Heathrow and the excellent internal communications with the rest of the UK, for both the import and the export of goods – a factor of increasing importance as the firm is becoming more marketing-led. The great majority of the professional people are highly skilled and experienced and come from other high tech companies; the firm is suffering from a national shortage of well-qualified people. Defence contracts play a significant role but the site is well placed for MOD contacts.

Firm 26 is the British distribution and service subsidiary of a Swedish firm which manufactures the cash dispensers used increasingly by banks and other financial institutions. It was located in Reading in 1979; the firm advertised for a managing director and the chosen person lived there. In 1982, after reorganization, the unit moved to other premises in the town. Heathrow, the motorway network and access to local component supplies are seen as the main locational advantages (though the machines themselves are shipped by sea); discussions with the bank headquarters are in London, though the service engineers operate nationwide. The unit has expanded to employ 42 people.

Firm 30 is the British sales, assembly and servicing subsidiary of an American firm specializing in a wide range of electronic equipment, especially in defence-linked avionics. The unit in an industrial park at Earley outside Reading has 150 employees. The original UK operation, in 1955, was at Staines; in 1964 came a move to Cranford near Heathrow, and in 1979 to Reading. Critical in this choice was the location of employees, access to Heathrow for both imports and exports, and availability of premises. The firm recruits its high-level professional staff mainly from inside the industry.

Firm 34 is the British subsidiary (sales, storage and administration) of a well-known American manufacturer of floppy and hard disks for computers. The company originated in Silicon Valley in 1972. It hired a British managing director in 1982 with a brief to set up the operation; his personal residential constraints limited the search to an area bounded by Newbury, High Wycombe, Southall and Basingstoke. Several premises were rejected as too small, too old or too restricted in use; the present Reading site was chosen as a self-contained unit with good access to the M4, good frontage for accommodating truck trailers, and with planning permission easily obtained. Originally the customers were highly concentrated in the South-East, but now they are nationwide. The unit now employs 45 people.

Firm 43 is the British subsidiary of a large US-based conglomerate multinational which contains specialized electronics interests. Located at Winnersh outside Reading in 1983, the unit (with 578 employees) specializes in coin mechanisms for vending machines; vigil radar for small craft and marine electronics; and test computers. Its history is a complex one. The parent company set up a British subsidiary at Slough in the 1930s to manufacture chocolate bars, the product for which it is still best known to the public; it diversified (among several other food-related activities) first into computer applications for market research and later into vending machines for the catering trade, out of which developed a specialized interest in coin-operated mechanisms. The Winnersh site was chosen because of its good road and airport links, for moving output within the UK and abroad, for obtaining supplies and for access to customers; its closeness to Slough, so the firm could transfer existing personnel; and the availability of the right kind of technical labour.

Firm 44 is the British subsidiary (parts sales, repairs and training) of an American manufacturer of disc drives and other computer ancillaries. It was located on its present Reading site in 1969. Critical factors were nearness to Heathrow, as this was to be the base for entry into the European market, coupled with lower rents than London.

Factors of location: a closer look

These case histories fortify the suggestion that emerged from the summary in Table 5.1: that the critical factors of location may be quite different for firms of different types. Table 5.2 takes this point up by analysing in greater detail the firms' replies on the relative importance of different location factors. However, because firms of Type 2 are so few in number, in order to obtain meaningful percentages we have had to aggregate them with Type 3 cases to produce a single category of multi-site firms.

The firms were asked what they considered to be the advantages of the eastern part of the 'M4 Corridor' (or the 'Thames Valley') compared to other parts of the country. This avoids the problem of ex-post rationalization by the interviewee when presented with a list of factors and asked which were important in the initial location decision. The results also suggest what characteristics should be present in, say, a development area, in order to attract such firms. The main factors of importance, as tabulated in Table 5.2, were labour, environment, communications, agglomeration characteristics and premises.

A pool of professional and skilled labour was often seen as a major attraction of the area: 20% of firms specifically stated that the availability of professional and scientific staff was an important advantage of being located in this area. While there was a shortage for many firms of many types of engineers and other professional workers, this was seen as a national, not local problem. Types 2 and 3 firms put a greater emphasis on the availability of professional workers (28%) compared to single-plant establishments (14%), although this may reflect the demand for larger numbers of such staff by the generally larger, more R&D or customer service oriented branch plants. The advantages of the area for obtaining skilled production and supervisory staff was about equally important (18%) for each type of firm. However, some 14% of single-site firms and 11% of branch plants (mostly manufacturing plants) considered it difficult to get skilled production workers in this area.

Studies of other high tech areas, such as Silicon Valley and Cambridge, have suggested that good environmental quality is a critical factor (Saxenian 1985a; Segal Quince 1985). In the Berkshire study some 30% of firms considered the environment to be an advantage of the area, although only two firms mentioned it as having any significance in their location choice. The good environment was generally expressed as 'a pleasant place to live', although 8% of firms said cultural/recreational facilities were also important. Of course while decision makers may perceive this to be a more pleasant place to live than elsewhere, it may be based on lack of knowledge of other areas. On the other hand 25% of firms considered high housing costs as a disadvantage of the area, often leading to difficulty in bringing in

Table 5.2 Perceived advantages/disadvantages of a Thames Valley location.

	Percentage of firms					
	Type 1 single-site		Type 2+3 multi-site		Total firms	
	−	+	−	+	−	+
Labour						
local availability of						
administrators & managers	−	−	−	17	−	8
professional & scientific staff	5	14	6	28	5	20
clerical staff	−	9	6	17	3	13
skilled & supervisory staff	14	18	11	17	13	18
semi- & unskilled staff	14	5	6	17	10	10
easy to attract labour	5	5	6	17	5	10
cost of labour	9	−	6	6	8	3
Environment						
housing cost	9	−	44	11	25	5
housing availability	9	−	−	17	5	8
cultural/recreational facilities	−	9	−	6	−	8
pleasant place to live	−	23	−	17	−	20
good environment (not specified)	−	5	−	22	−	13
social relations with others in the same industry	−	−	−	17	−	8
Communications						
Heathrow Airport	−	77	−	72	−	75
other airports	−	9	−	11	−	10
M4 motorway	−	73	−	50	−	63
other motorways and major roads	−	36	−	44	−	40
rail network	−	23	−	22	−	23
Agglomeration characteristics						
access to						
government research establishments	−	5	−	17	−	10
universities/higher education	−	−	−	11	−	5
local business services	−	5	−	11	−	11
local suppliers	−	23	−	6	−	18
local customers	−	36	−	39	−	40
exchange of ideas with others in the industry	−	−	−	6	−	3
near national government depts	−	−	−	11	−	5
close to other parts of firm	−	−	−	22	−	11
access to private R&D facilities	−	−	−	6	−	3
Premises						
suitable quality	5	23	−	44	3	33
suitable availability	5	41	−	39	3	40
suitable rent levels	18	9	11	17	15	13
suitable rate levels	9	−	6	6	8	3

Source: Interviews.
Note: − signifies disadvantage of area; + signifies advantage.

technical and skilled workers from other parts of the country.

Three-quarters of firms cited Heathrow as an important advantage of the area. Surprisingly, slightly more single-site firms saw it as advantageous (77%) than did multi-plant firms (72%). Other airports (only 10%) were not seen as very important. The M4 motorway was important for 63% of all firms (73% of single-site firms and 50% of multi-site ones), while some 40% of firms thought other major roads to be an important advantage, particularly for north–south travel. And – a somewhat unexpected finding – the railway was important for nearly a quarter of the firms (23%). The main reasons for the importance of air communications were the movement of output and inputs to and from abroad and movement of personnel (including a number frequently travelling to Scotland). The roads were especially important for moving output throughout the country (43% of firms) and particularly for site accessibility for customers (48% of firms). Generally road communications were more important and air movement of goods less important for single-site compared to multi-plant firms (despite Heathrow being more often cited by the former). While these questions referred to the general location (the eastern end of the M4 Corridor), road networks are likely to be much more important for local or site selection. For the future, the M25 could have the same significance in location as the M4 had in the past – a point to which we return in Chapter 12.

By far the most important agglomeration advantage of the area, as reported in the interviews, was access to local customers (40% of firms). This was equally important for single- and multi-plant firms. For small firms, access to local suppliers was quite important (23% of firms), though this was less so for multi-site firms (6%) who received supplies from other branches of the company or who expected the suppliers to come to them (possibly reflecting the large purchasing power of many of these firms). For a number of the multi-branch firms (22%) the location of other parts of the firm in the UK was important. Surprisingly, access to business services was considered advantageous in this area by only 11% of firms, perhaps as they were perceived to be readily available in most parts of the UK. Access to local universities – important in Cambridge and Silicon Valley – were thought important by only 5% of firms. This helps to indicate the different nature of high tech development in the M4 Corridor compared to those other areas.

About 40% of both single- and multi-plant firms reported that suitable premises were an important factor – but at a price: 18% of single-site and 11% of multi-site plants thought the area disadvantageous in terms of rent levels. Rate levels in the area were thought excessive by only 8% of all firms. Of course, rate levels are considerably below those in Central London (Anon. 1984) – and this may be the standard of comparison.

We also asked about the factors that might be helping or hindering the expansion of firms (Table 5.3). Virtually the only important factor

Table 5.3 Factors aiding or hindering the expansion of businesses.

| | Percentage of firms | | | | | |
| | Type 1 single-site | | Type 2+3 multi-site | | Total firms | |
	+	−	+	−	+	−
demand for output	36	0	33	0	35	0
labour shortage	0	36	0	22	0	30
labour costs	5	5	6	6	5	5
working practices	0	0	6	6	3	3
financing constraints	5	18	0	22	3	20
site constraints	0	27	0	11	0	20
no answer	27		50		38	
(number of firms)	(22)		(18)		(40)	

Source: Interviews.

helping expansion was increases in demand for the firm's output. This was mentioned by 36% of single-site and 33% of multi-plant firms. Conversely 36% of single-site and 22% of multi-plant firms were hindered by labour shortages, and only 5% overall by labour costs. When firms not answering this question are excluded, the labour shortage figures are 50% and 44% respectively. The two other main hindrances to firms hoping to expand were site constraints (27% of single- and 11% of multi-site firms) and financing constraints (18% and 22% of firms respectively). Interestingly, only one firm considered working practices a hindrance. In general, then, expansion depended upon rising demand, and was hindered by financing constraints, site constraints (for single-site firms) and labour shortages, but not usually by labour costs.

Towards a synthesis: how the complex grew

This analysis helps us to develop hypotheses about a theory of location for firms in the Western Crescent, which we shall go on to develop in Chapter 6. We find that there are just a few older-established firms, though one or two of these were major employers of labour. Relatively few firms had been founded in London. But the really significant factor appeared to be breakaway or spin-off from established firms, some of which (Plessey, EMI) had a London origin. And another major source of growth was the multinationals – particularly the Americans – which were drawn to the area because of its easy access to Heathrow and to the British market, and because of its established supplies of high-quality technical labour.

This suggests some hypotheses. London's unique concentration of high technology industry perhaps provided one origin of the Western Crescent – but mainly at one remove, through spin-off formations of new companies by ex-employees. Some – but, it should be emphasized, a minority – were also pulled by the location of the GREs. And, as the process of spin-off growth occurred, the resulting concentration of specialized labour proved attractive to companies from abroad seeking to establish a British or European base. For them, however, the critical factor was the conjunction of these labour supplies with the unique international access offered by the airport and its motorway connections.

The historical curiosity, then, is that a set of contingent factors seems to have worked in the same direction, thus producing a virtuous circle of cumulative and self-reinforcing growth. In Chapter 6, we shall try to develop an eclectic locational theory around these speculations; then, in Chapters 7–11, we shall examine the major factors separately and in detail.

PART III Why?

6

Towards explanation: the geography of innovation

Having established the facts of high tech location, and having asked the firms themselves about their locational decisions, we now turn to the question: why? Why, in contemporary Britain, should the sunrise industries concentrate so strongly in the Western Crescent and the Western Corridor? Why should they avoid – with a few significant exceptions – the older industrial areas of northern and western Britain? And what explains these few exceptions: the Greater Manchester area, the Silicon Glen area of central Scotland? Is there a reason why such industry should turn its back on the inherited industrial traditions and skills of the past, preferring greenfield locations in the small country towns of southern England? Is the reason to be sought in the relation of the Crescent – and the Corridor – to London? And, if so, what is the source of London's magnetic pull? These are the kinds of questions that we shall seek to answer in this central part of the book.

To do so, we need to start by searching for clues in the theory of location. Economists and economic geographers, over a century and more, have developed a large theoretical literature to try to explain why economic activities locate where they do. Until the mid-1970s, most of this literature was in the neo-classical tradition of economics: it sought to explain location – of firms, of whole industries, of towns, of land uses – in terms of spatial equilibrium, reached at a particular point of time in response to different locational pressures. During the late 1970s and the 1980s, a different tradition has arisen, based on concepts of political economy: it seeks to explain location in terms of changing structural conditions in the capitalist economy, viewing this as a worldwide system in which individual firms seek to preserve or enhance their position in the face of competitive pressures. Whereas the old tradition dealt in static, point-of-time relationships, the new tradition focuses on the dynamic – even unstable – character of the modern capitalist economy.

Neither of these bodies of theory, as yet, provides the vital clue that would solve the mystery of the high tech phenomenon. It is necessary to borrow pieces from different parts of theory, to try to build an eclectic explanation that will help (Markusen *et al.* 1986). To do that, in relation to the observed facts of high tech location in Britain in the 1980s, is the task of this chapter.

Neo-classical explanations

First, there is a well-established and well-known body of traditional location theory in the neo-classical tradition, building on the foundations of von Thünen (1966 (1826)) for agriculture, Weber (1929 (1909)) and Hoover (1948) for manufacturing and Christaller (1966 (1933)) for services. These are integrated in the general location theory of Lösch (1954 (1944)). Unfortunately, only parts are relevant for understanding the high tech phenomenon.

Weber's theory is based first on a so-called locational triangle, at the corners of which stand raw materials, labour and markets. Seeking a location somewhere within this triangle, an individual firm (and, through the aggregation of individual decisions by firms, a whole industry) will make successive adjustments until it finds the lowest-cost location, as determined by the relative weights and transportation costs of the three factors. The initial location, thus determined, will be affected by the existence of agglomeration economies, which create a locational inertia for certain interrelated industries: they stay in their initial location because of the advantages they get from being close to each other and to the general network of services that the location provides. In the more general Löschian model, these centres of activity serve wide surrounding hinterlands, producing an urban hierarchy in which more specialized functions – and also those with greater economies of scale – concentrate in the higher-order cities.

Applied to high tech industry, some parts of this traditional theory seem hardly relevant, while others are very much so. Both materials and product are extremely valuable in relation to bulk and weight; transportation costs are minimal; the product may move long distances, often by air, during various stages of fabrication; speed and promptness of delivery may be more important than ton–mile costs. So air travel is important, but there is a chicken-and-egg question here: the electronics industry, growing up with the air age, became multinational in its organization and its markets, but it is impossible to say whether this led to the use of air cargo or was caused by it. But, because of the tradition of holding small stocks and being quick on to the market with the new product, speed rather than cost of freight becomes the critical consideration.

Labour is more important for high tech, but in two different ways: typically, high tech industries have sharply bipolar labour markets, with small numbers of skilled professionals and a much larger number of relatively low-paid process and assembly workers (Saxenian 1985a). The first, it is surmised, are highly mobile and need to be attracted by good working and living conditions in the form of good climate, good housing in a pleasant environment, and good cultural and educational facilities; thus firms may follow their key workers, rather than the other way round (Berry 1970, pp. 40–1). The second, in contrast, will be found in low-wage, weakly unionized, rural locations. These two kinds of location may overlap, but probably will not; therefore, the argument runs, the industry may split locationally into two components, with research and development in the first, routine manufacture in the second.

The final element of Weberian theory, agglomeration economies, may be very significant for high tech. Such industry is typically very innovative and – at least in the early stages – is dominated by small firms, which will tend to depend on external economies: access to skilled labour, to specialized services, and to technical information. Markusen *et al.* (1986) suggest that an area like Silicon Valley bears an extraordinary resemblance to the Victorian industrial quarter as described a century ago by Alfred Marshall:

> When an industry has chosen a locality for itself, it is likely to stay there long: so great are the advantages which people following the same skilled trade get from near neighbourhood to one another. The mysteries of the trade bcome no mysteries; but are as it were in the air, and children learn many of them unconsciously. Good work is rightly appreciated, inventions and improvements in machinery, in processes and the general organization of the business have their merits promptly dis-cussed; if one man takes up a new idea, it is taken up by others and combined with suggestions of their own; and thus it becomes the source of further new ideas. And presently subsidiary trades grow up in the neighbourhood, supplying it with instruments and materials, organizing its traffic, and in many ways conducing to the economy of its material. (Marshall 1920 (1890), p. 225)

Detailed study of the growth of high tech regions shows just how striking these parallels are. Thus, in Silicon Valley during the 1950s, one firm spawned another; one alone, Fairchild, gave rise to no fewer than 50 (Saxenian 1985a, p. 25). (In the Scottish electronics industry after World War II, Ferranti seem to have played a similar role.) The majority of Silicon Valley engineers seem to have been recruited locally, either from the Stanford or Berkeley campuses or by firms' own recruitment efforts; although larger firms have recruited from outside, once in the area skilled workers stay because job prospects are better. And computer manufacturers were drawn to the place where semiconductors were made (Markusen *et al.* 1986). As such a complex grows, linkages between firms – one providing parts or subassemblies or specialized processes for another – become ever closer and more complex, tying all of them together.

This, however, leaves a critical puzzle, which traditional theory does not adequately address. In Marshall's time and even much later, such agglomeration economies characteristically operated near the heart of the city, in areas such as the garment quarters of London and New York (Helfgott 1959, Hall 1962, ch. 4, Martin 1966, ch. 7) or the jewellery and gun quarters of Birmingham (Wise 1949). Yet today's high tech industry seems to eschew such locations, preferring areas that were not traditionally industrialized; the industrial seedbed function of the inner city has been lost (Hall 1981, p. 42). As Massey (1984) and others have suggested, this may be because the crucial innovative group, having acquired mobility through car ownership, have moved their homes out of the city and into high-amenity countryside. And, in an age of international telephone hook-ups and technical journals, traditional city-centre face-to-face contact may be far less important than it was. Or, perhaps, it takes a different form: no longer chance encounters in the city street, but technical conversations in restaurants and bars and in gatherings like Silicon Valley's Home Brew Club, where the personal computer evolved (Freiberger & Swaine 1984, ch. 4). But there is still a mystery: why not any high-quality rural area? And, if face-to-face is less important, why the continuing tendency of high tech to agglomerate?

Political economy theory

This suggests a central weakness of traditional theory: that, dealing essentially with static equilibria, it cannot account for the character-istically dynamic, unstable capitalist economy. Clearly, such a weak-ness will be very significant for the newest, most dynamic parts of the industrial economy. Any adequate theory must explain the creation of these new innovative enterprises, their subsequent growth and possible locational shift.

It is just this dissatisfaction that has led so many contemporary economic geographers to embrace theories of political economy. Working within a marxist framework, they argue that:

> if . . . geographical patterns are the outcome of socio-economic processes (operating over space) then in order to understand a pattern we must go behind it and *interpret* it in terms of the structures and processes on which it is based. (Massey 1984, p. 67)

Specifically, they argue that the modern capitalist corporation seeks constantly to rearrange the spatial division of its activities so as to reap maximum profits. Internationally, the multinational corporation views virtually the whole world as its own space economy which it can manipulate to its advantage – typically by a hierarchical division of activities, with overall control maintained in the parent country and city, divisional headquarters in other developed countries, and production in low-wage developing countries. Nationally, the multi-plant company has its headquarters in the central metropolitan region, branch plants for routine production in lower-wage, government-assisted peripheral regions (Table 6.1). Multinationals may operate an A–B–C structure in Hymer's (1975) model; smaller national corporations a B–C structure; very small single-plant companies simply have a C structure. Applied to the United Kingdom, this implies head-quarters offices – main ones for British companies, divisional for the multinationals – in the South-East, branch plant production in regions such as Scotland and Wales. Naturally, in such a pattern, peripheral regions have little control over their economic fates; the crucial decisions are taken hundreds or thousands of miles away.

Recent work suggests that high technology companies may play a particular role in such structures (Massey 1984, Castells 1985, Sayer & Morgan 1987, Henderson & Scott 1987). Because they are new, they are still developing their global strategies. But because they are high tech, they can more readily spread activities across the globe while retaining close control from headquarters. This poses the question: does the Western Crescent perform an A role in Hymer's conception, serving as a global headquarters for British multinationals? Or is its role a B and D one, acting as a divisional headquarters for overseas companies or as

Table 6.1 Scheme of corporate control.

Level of corporate control	Type of area		
	Major metropolis (e.g. New York)	Regional capital (e.g. Brussels)	Periphery (e.g. Ireland, South Korea)
1 long-term strategic planning	A		
2 management of divisions	D	B	
3 production, routine work	F	E	C

Source: Hymer 1975.

headquarters for smaller British firms? Or is it actually at the lowest level of the hierarchy, acting as routine production location for firms headquartered elsewhere and as home for small single-location firms?

Sayer (1985) argues that there is a danger in relying too heavily on models of corporate activity of the type produced by Hymer (1975). He contends that corporate spatial organization, and general corporate practices, will always be to some degree unique, and hence will not fit into neat hierarchical conceptions. An interesting example in Berkshire of just such an exception is the major electronics company, Racal. The company is organized into a large number of mainly autonomous units, each with a Racal-suffix name. Each unit tends to perform all corporate functions – R&D, production, marketing and so on – from one or a small number of closely located sites. This company structure can be fitted into Hymer's model, possibly on an A–D–F basis, but is not consistent with standard conceptions of corporate spatial hierarchies.

Massey (1984), in a recent study, argues that in the electronics industry there are actually two hierarchies: a managerial hierarchy of ownership and control, which distinguishes headquarters from branch plants; and a production hierarchy, which separates R&D from the production of complex components, and that in turn from final assembly, each with a different kind of labour and a different degree of autonomy in the labour process (ibid., pp. 70–3). Often, though not inevitably, these different functions were geographically separated:

> At one end of the scale the celebration of individualism, at the other dispensability and infinite replaceability. It is a social contrast with definite spatial coordinates. It doesn't just stretch, it is deliberately spaced, from Palo Alto, California, to the Masan Free Production Zone, South Korea. (ibid., p. 137)

And the British industry, with its concentration of R&D staff in the South-East, well illustrates a mutually reinforcing relation between location and social status: industries locate here, between the Solent and the Cotswolds, because this is where the right kind of people are (ibid., pp. 141–2). Large firms will separate this kind of activity from low-paid assembly operations; smaller firms, producing in small batches, will tend to cluster all their operations in the South-East where the necessary quality of staff is more readily available (ibid., pp. 150–3). Thus, Massey argues in another contribution, the prime differentiators between regions are no longer based on industrial structure but on occupational structure, both within and between industries (Massey 1985, p. 306).

But this assumes the existence of both the firms and the labour force; it does not explain how they come into being. The nearest Massey comes to this is in a reference to small companies that have entrepreneurship, growth-potential and enterprise; they tend to be small and relatively new, and they are highly concentrated in the 'sunbelt' from East Anglia to Southampton and Bristol. This, she explains, is because the sectors concerned – electronics, instrument engineering, producer services – are there, as is the spectrum of the population likely to start new firms, with access to capital, previous management experience, education and local markets (ibid., pp. 281–2). But in all this, there is an element of circularity. Perhaps, indeed, success breeds success. But what started the area and its people on this road?

There are some clues in the American work of David Birch (1979). He suggests that everywhere the capitalist economy generates large numbers of new firms, but many die. The point is that death rates do not vary much from place to place; birth rates, and expansion rates for existing firms, do. They are higher in America's growth region – the Sunbelt of the South and West – than in the declining Frostbelt. Bluestone and Harrison, in their political economy analysis of the deindustrialization of America, quote Birch's work extensively to illustrate the fact that many capitalist firms die – but the fact remains that in the vigorous regions, the birth rate is even higher than the death rate (Bluestone & Harrison 1982, pp. 29–31). And this points to a central weakness in the analysis of the political economy school: their theory is better at explaining plant deaths and job losses than plant births and job creation. It is no accident that the classics in the *genre* – Massey and Meegan on *The anatomy of job loss* (1982), Bluestone and Harrison on *The deindustrialization of America* (1982) – have focused on deaths and losses; even though, in the latter case, the statistical evidence in the book shows clearly that in the United States between 1969 and 1976, there was large-scale net job *creation*. So there is a mystery here, and it remains unsolved.

Product and profit cycles

Birch does not provide an explanation, either; but part of one can be found in the concept of the product cycle, as first enunciated by Raymond Vernon (1966) and as developed by Ann Markusen (1985). The argument here is that products – and the firms and industries that make them – go through cycles from youth through maturity to old age. As they do so, they demonstrate striking changes in growth rate, profitability, degree of concentration, and location.

In infancy, a new industrial sector is preoccupied with designing and commercializing a new product: it consists of a number of new, small firms, dominated by skilled engineers, technicians and other designers, often producing in small batches to special orders, and earning high profits. These firms classically agglomerate in Marshallian fashion. Later, in youth, firms become centrally concerned with mass production and market penetration; growth rates and profit margins fall somewhat. At this stage, more standardized production processes are spun off to lower-cost locations outside the original agglomerate core, but generally within the same region. Then, in maturity, the market becomes saturated and competition is increasingly fierce; firms compete either by cost-cutting or by collusive agreements to share the market. Now, firms increasingly decentralize production to lowest-cost locations, often in developing countries or depressed older industrial regions with surplus labour. Finally, in old age, competition drives the price of the product below the accepted rate of return; firms close down, or ruthlessly rationalize to cut costs. Generally, the oldest and least efficient plants in the original core regions will be closed in favour of newer factories established in the previous decentralization phase; though in a few cases (as currently in the American car industry), remote branch plant operations may be closed in favour of the central headquarters complex (Markusen 1985).

Product-profit cycle theory thus provides a complete explanation of why industrial location should first concentrate in a core region, later decentralize to neighbouring locations, and still later spread world-wide. However, what still remains elusive is the basis of the original industrial core. Some writers have suggested that the majority of new firms simply spring up in founders' home towns; but, more often, the success of a new entrepreneur is related to institutional advantages which in effect produce instant agglomeration economies (Markusen *et al.* 1986). Thus in the United States, the two most celebrated high tech areas – Highway 128 around Boston and Silicon Valley – clearly owed much to spin-off from research at MIT and Stanford respectively (Saxenian 1985a,); while military and space research has powerfully aided the development of major centres such as Houston, Huntsville (Alabama) and Melbourne–Titusville (Florida).

Towards a geography of innovation

To try to generalize from these cases, so as to develop a general theory of initial high tech location, we can borrow from the Schumpeterian concept of innovation (Schumpeter 1961 (1911), 1939). Innovation, for Schumpeter, was synonymous with entrepreneurship or enterprise; it was the central feature of the capitalist system. It consisted not in invention, but in the establishment of a new product or technique or method of production in the actual market; it could, importantly, include the application of an invention in the form of a commercially marketable product. Indeed, Schumpeter argued, the history of capitalism had been marked by a series of long waves of development – first identified by the Soviet economist Kondratieff (1935) – each of which represented a product cycle triggered off by a technological innovation. The first, from approximately 1785 to 1842, had been based on cotton textiles, the smelting and refining of iron, and the stationary steam engine; the second, from 1842 to 1897, on railways and Bessemer steel; the third, from 1897 onward, on the chemical, electrical and motor vehicle industries (Hall 1985, p. 6).

Much criticized on its first appearance and afterwards, especially in the work of Kuznets (1940, 1946, 1966), Kondratieff–Schumpeter long-wave theory has enjoyed a revival during the recession of the late 1970s and early 1980s – not least because the theory predicts precisely such an event (Mandel 1975, 1980, Rothwell & Zegveld 1981, Rothwell 1982, van Duijn 1983). In fact, most students of the subject now seem to accept the existence of long waves of 50–60 years in duration; but they disagree about the cause. Some argue that clusters of innovations are indeed the origin (Mensch 1979); others argue that exogenous, often purely contingent forces are responsible (Freeman *et al.* 1982, Clark *et al.* 1984). Long-wave theory is, however, completely aspatial; it does not suggest why innovation should cluster at one time in one place, at another time in another. But it is not difficult to cull elements of a location theory of innovation from the economic history literature. Many studies, some now ancient, suggest that innovations seldom develop in older-established industrial locations, which develop some kind of hardening of the innovative arteries. Thus the classic early studies of the English woollen industry suggest that it moved first to Suffolk, then to Norfolk, then to the West Riding of Yorkshire, as the previous centre failed to adopt technical innovations (Clapham 1910, Carus-Wilson 1941); Checkland (1975) argues that Glasgow's industrial decline resulted from its obsession with custom-built ships and its failure to develop new industries like vehicles and aircraft; Chinitz (1960) and Markusen (1985) in the United States, and Fothergill and Gudgin (1982) in Britain, suggest that domination of local resource markets by big industries explains the poor growth records of places like Pittsburgh, Detroit and Derby; while in contrast, Allen's classic work on Birmingham (1929) indicates that the city's success lay in the continuing innovative ability of its many small firms – a tradition that may now have been lost through industrial concentration and dominance by the car industry.

The positive characteristics of an innovative environment, however, remain shadowy. It is unlikely to occur in an older industrial region – though, *vide* Allen's study of Birmingham, even that is not certain. It may be helped by a positive business climate, with not too much regulation and with low rates of taxation – as suggested by advocates of the Enterprise Zone idea (Butler 1981) and by students of the American Sunbelt (Perry & Watkins 1978). But this somewhat ignores the fact that in the modern capitalist economy, much innovation arises from organized research and development which is increasingly concentrated in large laboratories, both public (universities, government research establishments) and private. Hence the argument that the geography of innovation follows the geography of R&D.

Fortunately, an abundant literature on this subject exists, both from the United States (Malecki 1980a, b, c, 1981a, b) and from Great Britain

(Buswell & Lewis 1970, Howells 1984). Malecki suggests that there is a fundamental distinction between industrial and government-based R&D. The first tends to be concentrated in the older American manufacturing belt, some of it in headquarters of major industrial corporations (Pittsburgh, Detroit), some of it in non-headquarters industrial cities. The second, in contrast, is mainly in university cities (Austin, Texas; Lincoln, Nebraska) or centres of federal government research (Washington, DC; Huntsville, Alabama), many outside the manufacturing belt.

It is this second type, we could argue, that is truly innovatory in the sense that it can actually generate new industry. Most of it is defence- and space-related and is geographically highly concentrated: no fewer than 61% of all aerospace R&D laboratories are in the Los Angeles area, while much of the electronics research is found in only five areas: Boston, New York, Philadelphia, San Francisco and Los Angeles (Malecki 1981a).

Similarly, British studies show that R&D is heavily concentrated in the South-East region around London, especially west of the capital where the government research establishments tend to be grouped, and where much of the new high tech industry has developed. Indeed, the work of Oakey *et al.* (1980) and of Thwaites (1982) shows just how far the South-East and its neighbouring regions lead the rest of the country in their rate of industrial innovation; in contrast, older-established industrial regions – Wales, Scotland, the North-West and West Midlands – tend to perform relatively poorly.

This evidence receives strong confirmation from more detailed local case studies, such as those by Saxenian (1985a, b) and Rogers and Larson (1984) of the development of Silicon Valley and Boston's Highway 128, or that by Feldman (1985) on the incipient biotechnology industry in the San Francisco Bay Area. All these suggest that the presence of local university R&D was of critical importance in the supply of new graduates who established new, innovative firms, and who then continued to draw on the stream of university research and on the supply of new scientists and technologists.

An eclectic explanation

If we now try to put these theoretical pieces together, the result is an eclectic account of high tech locational evolution which reads as follows.

Approximately every 55 years, a bunching of technological innovations will help to produce a new batch of fledgling high tech industries. (Scholars differ on whether the innovations themselves are the primary cause, or whether some other mechanism works to produce them.) Each of these innovation swarms is likely to arise in a location different from the previous one, because older-established industries create what Checkland (1975) calls a upas tree effect: their shade kills the development of new innovative impulses. Very big cities with diverse industrial traditions, especially major capitals with their small craft industries, may prove exceptions to this rule, particularly in the earlier long waves when much innovation resulted from casual experiment by individual technologist-inventors.

However, in the twentieth century, systematic R&D in organized laboratories – both public and private – becomes much more important for the innovation process. Since much industrial R&D is concerned with the improvement of existing products or the development of new products closely related to old ones, really innovative R&D – the kind that actually creates new industries – is increasingly likely to occur in public laboratories, either in universities or in government research establishments. Hence the location of such research is likely to prove critical for the development of high tech.

Once established, the newly founded industries go through the product–profit cycle. As they do so, they will decentralize parts of production first to branch plants located fairly close to the parent establishment, then – in later stages, as competition becomes fiercer and profits fall – to more distant locations, including offshore operations. In the twentieth century, because of the development of transnational capitalist firms, this process becomes ever more common. But generally, only the routine manufacturing functions will be so transferred; research and development, together with the ultimate control of production and sales, will remain in the original headquarters location.

It is also conceivable that events in one such cycle may indirectly influence events in the next. Some major industries may continue to grow through two or more cycles, experiencing major shifts of world location as they do so (see, for instance, Altshuler *et al.* (1984, ch. 2) on the automobile industry). Some industries, arising in one profit cycle, may spawn others in the next – as, during the current long wave, the electrical industry has spawned the electronics industry. In these circumstances, the location of the parent industry may influence the location of the offspring. But, since the parent industry will itself have experienced a shift of location during its own profit cycle, the resulting locus of the new industry is likely to be different from the original locus of the old.

It may, however, not be very distant. The initial stage of decentralization will usually take a fairly new, still fast-growing firm to a location in the same region, either in the form of a branch plant or in the form of complete relocation, including R&D facilities. These facilities may well then provide one basis of spin-off of new firms, including firms in new activities at the start of the next product cycle. In these circumstances the new industrial complex would develop only a few miles from the old, probably in a peripheral location at the edge of a major urban area; thus Silicon Valley and Highway 128 at the peripheries of their respective areas.

The major complicating feature in this process is the injection of organized, publicly financed R&D in the most recent product cycle – which, it is hypothesized, began shortly after World War II. About this time, older-established universities greatly expanded their programmes of scientific research; some new universities were founded, though probably few developed quickly as major centres of research; and, perhaps most importantly, new government-financed laboratories were established, predominantly in the military and space fields. The location of such facilities, and the reasons for location, therefore become critical. Where existing facilities were extended through government contracts, as at Stanford University, or where such existing facilities spawned nearby new ones, as in the relationship between Berkeley's Lawrence Berkeley Laboratories and the Lawrence Livermore Laboratories some 20 miles (32 km) distant, clearly locational inertia played a powerful role. However, where completely new facilities were established, as in the case of some of Britain's government research establishments between 1935 and 1946, then the reasons for location may prove to be of crucial importance.

Three working hypotheses

In the chapters that follow, we shall seek to test different elements of this eclectic theory. In particular, we shall examine three working hypotheses that stem from it.

First, the *local displacement* hypothesis: that any one generation of high technology industry will locally decentralize in the course of the profit cycle, thus providing a new locus for the next generation of high tech industry that it may spawn. This suggests that new high tech locations will tend not to be identical to old ones, but to be close to them. We shall examine this hypothesis, applied to the case of the Corridor and Crescent, in Chapter 7.

Secondly, the *organized research* hypothesis: that, especially in the product cycle after World War II, the location of major centres of government-financed R&D has been crucially important in the growth of well-established, large high tech firms and in the generation of small, new, innovative ones. We shall examine the growth of government research in Chapter 8 and its linkages with industry in Chapter 9.

Thirdly, the *international capital* hypothesis: that, again especially in the most recent cycle, successful high tech firms have increasingly become transnational corporations, placing their productive activities on a worldwide basis, and responding to competition by choosing the lowest-cost locations where effective production is possible. Notice that this does not mean the lowest-cost location regardless of all other considerations; since high tech industry has particular requirements in terms of international access, labour supplies and specialized services, the resulting location will inevitably be a compromise.

In particular, multinationals from advanced, high-wage economies (the United States, Switzerland, Sweden) may establish operations in Britain to obtain the advantages of cheap (and well-qualified) labour and to establish a European marketing and servicing base. They will tend to establish such a base in the existing area of industrial concentration, with particular reference to international communications, implying a location in the M4 Corridor close to Heathrow. Since they are large, they may well decentralize routine assembly operations to lower-wage peripheral regions. We will examine some of the implications of this in considering transport factors, in Chapter 11, and return to it when we look at the response of the firms themselves, in Chapter 12. These, then, are the three hypotheses that we shall test in the context of the Western Crescent and the Western Corridor.

7

The origins

Perhaps – as argued in Chapter 4 – Britain's high tech core is a crescent; perhaps that crescent contains a small but fast-growing corridor. Whatever the geometric analogy, the area under study is a relatively recent feature of the British economic landscape; it is a creation of the 1960s and, particularly, the 1970s. But its origins go much farther back than that, to the period between the two world wars and even to the nineteenth century. In those relatively distant decades, processes began to operate, and critical decisions were taken, which were to have profound consequences in our own time. This chapter and the one that follows attempts to tease out these long and sometimes obscure connections.

The story they tell is woven around two themes, which virtually provide alternative explanations of the origin of the Crescent. One, the theme of this chapter, is the steady – indeed, apparently almost inexorable – outward movement of the vanguard of the technologically advanced industries from their birthplace in London. The other, the theme of Chapter 8, is the concentration in the Crescent – beginning in the late 19th century, but accelerating remarkably during and just after World War II – of the complex of defence and related government research establishments, which seem to have provided much of the impetus to the development of high technology industry since 1940. Though these two processes provide alternative explanations, it could be that both are in part true; a judgement on that must wait until the end of the book.

The process of decentralization:
an overview

A graphic picture of the process of outward movement of high technology industry is provided by Table 7.1 and Figure 7.1, which are based on Census of Population data for employment (by workplace) for the years 1921, 1951 and 1971. Any comparison over such a length of time suffers from a basic difficulty: that many of today's high tech industries did not exist, or at least were not separately recognized, in the past. The 1951 Census did not distinguish a computer industry; the 1921 Census did not notice a radio industry. So, in order to get any comparison, it is necessary to take a wide definition – wider than anyone would use for recent analysis. In comparison with both the long and the short lists used in Chapter 2, Table 7.1 and Figure 7.1 include the whole of the electrical engineering industry, musical and scientific instrument making, and the manufacture of aircraft and their parts.

They reveal a clear, steady, outward trend. Inner London – the old London County Council area – contained over 27% of all high tech

Table 7.1 High tech industries (extended definition): employment 1921, 1951 and 1971.

'Old' county	Employment			%England and Wales			Location quotient		
	1921	1951	1971	1921	1951	1971	1921	1951	1971
England & Wales	196 710	897 994	1 103 100	–	–	–	–	–	–
Greater London	87 534	265 778	211 050	44.5	29.6	19.1	2.3	1.4	1.0
LCC	53 385	89 494	45 660	27.1	10.0	4.1	NA	1.3	0.4
Middlesex	19 433	106 952	88 670	9.9	11.9	8.0	NA	2.3	1.8
Surrey	2 258	34 899	50 940	1.1	3.9	4.6	NA	1.7	1.4
Hants	3 186	34 312	45 280	1.6	3.8	4.1	NA	1.5	1.3
Berks	129	5 062	16 660	0.1	0.6	1.5	NA	0.6	1.3
Bucks	118	10 495	15 040	0.1	1.2	1.4	NA	1.3	1.2
Oxfords	79	3 942	2 870	0.0	0.4	0.3	NA	0.6	0.4
Herts	1 578	26 450	50 950	0.8	2.9	4.6	NA	2.2	2.6
Beds	1 642	7 519	14 270	0.8	0.8	1.3	NA	1.0	1.3
Total, London NW Sector	81 808	319 125	330 340	41.6	35.6	29.9	NA	1.3	1.1

Sources: Census 1921, 1951, Industry Tables; Census 1971, Economic Activity Tables. 1971 figures for Greater London, Middlesex, Surrey and Herts are affected by boundary changes.

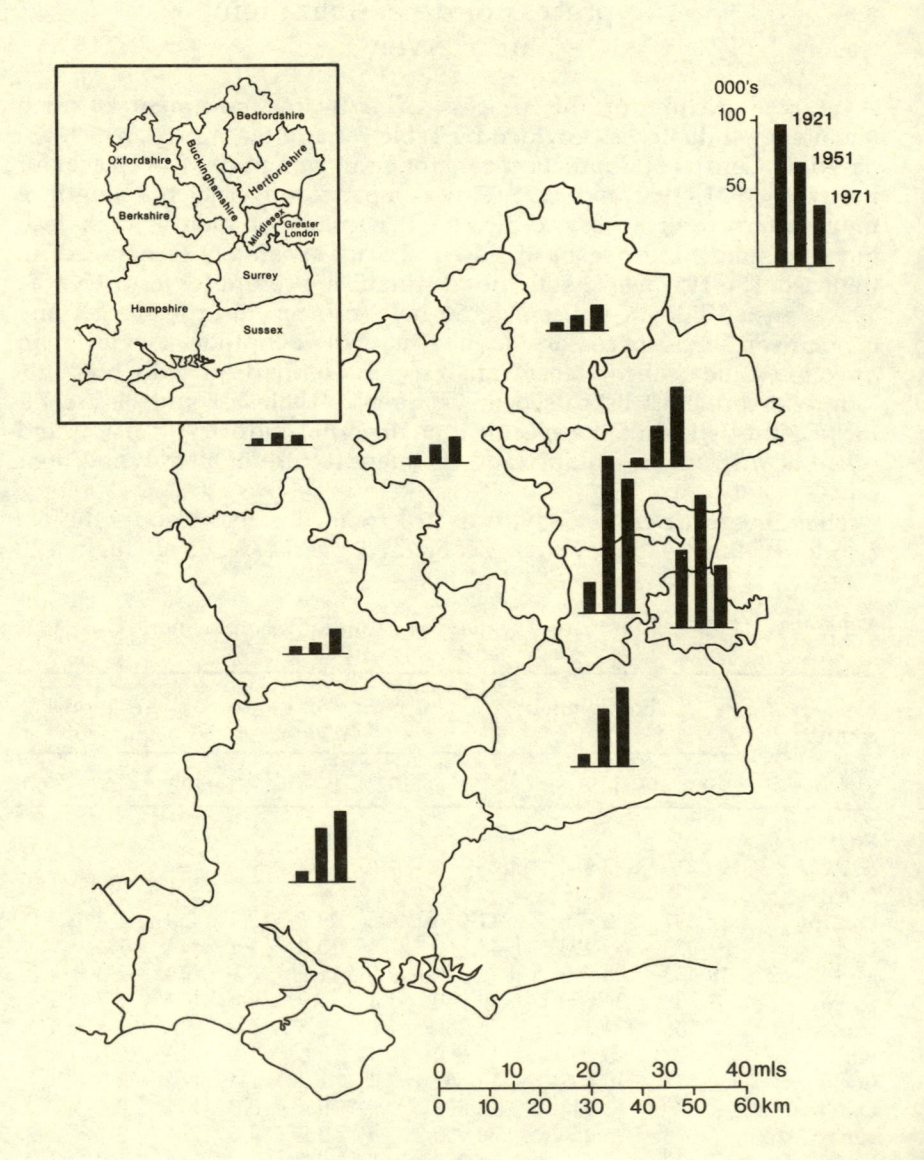

Figure 7.1 High technology industries (extended definition): employment (by 'old' counties), 1921, 1951 and 1971.
Source: censuses.

industry, thus defined, in England and Wales in 1921; that proportion had already sunk to 10% by 1951, and to a mere 4% by 1971. Greater London had no less than 44% of the total in 1921, retained nearly 30% in 1951, but had less than 20% by 1971. The counterpart of this decline is the growth of employment to counties that ring London to the north and west.

What is equally evident, though, is that this growth is quite recent. Between 1921 and 1951 – that is during the interwar, wartime and immediate postwar periods – the six outer counties of the sector increased their share from a mere 3.4% of total England and Wales high tech employment to 9.7%: an impressive growth in percentage terms, but still leaving them relatively insignificant compared with the major industrial areas. By 1971 they had jumped to 13.2%. But the *relative* concentration of high technology industry in the local economy, as measured by the location quotient, did not notably increase between 1951 and 1971; indeed, in a number of counties of the outer ring – Hampshire, Buckinghamshire and Oxfordshire – it actually fell. The striking exception, significantly, is Berkshire, whose location quotient more than doubled in these two decades. During this period, Greater London was actually losing high tech jobs.

Put another way, the dominant feature of the interwar and wartime period was the short-distance migration of high tech employment from inner to outer London, particularly to Middlesex, which saw a more than fivefold increase in high tech jobs between 1921 and 1951. From 1951 to 1971, as Middlesex too began to lose employment – a loss of more than 20% in 20 years – the beneficiaries were the outer counties of the Western Crescent. The wave of outward movement had simply rolled on.

The early origins: London's precision trades

'Movement' in this sense, of course, is merely a statistical artefact; the Census figures do not say how many firms and how many jobs actually moved out. But other accounts fill this gap. The origins go back into the 19th century and deep into what is now central London. Then, 'Certain quarters, of which Clerkenwell is the chief, played a crucial role in the development of London's modern precision and electrical industries' (Martin 1966, p. 35). Here Negretti and Zambra began thermometer manufacture in 1850, evolving eventually into manufacturers of aeronautical and automatic control instruments; here also, in Hatton Garden in 1883, Ferranti set up in business at the age of 19. Scientific glassware, an important adjunct to these trades, was another industry concentrated here: the firm of Cossor started in 1896 in Farringdon Road. During the 20th century this zone spread out in an arc a few miles to the north, from Dalston to Camden Town (ibid., p. 36).

Another origin of precision trades was Pimlico and Lambeth, which in turn engendered an industrial sector to the west – including, notably, the GEC electric lamp factory at Brook Green, Hammersmith, established in 1893 (ibid., p. 37). When large-scale manufacture of radio valves began in World War I, it mainly went to existing electric light manufacturers such as GEC; Captain Mullard's infant valve works, catering to Admiralty orders, was close by in Hammersmith. The Sperry gyroscope works – a subsidiary of the American firm – began in Pimlico in 1913, catering for Admiralty orders; as it expanded it moved, first to Shepherds Bush and then to the Great West Road at Brentford (ibid., p. 38).

It seems clear, then, that there was a continued process of development and movement. Beginning either in Clerkenwell or in Pimlico–Lambeth, new firms – the high technology firms of their day – moved out in search of space. In turn further new entrants established themselves in the new locations (such as Hammersmith), perhaps because they wished to tap specialist labour pools. Then these locations proved constricting, and firms moved out again – especially, just after the World War I boom in electronics, to the Great West Road. It is this process of movement that has played a critical role in the genesis and growth of the M4 Corridor.

The beginnings of the process can be traced as early as the 1890s, when Charles Booth's monumental survey of London life and labour reported of the electrical industries that 'No industry is more modern, nor does any show more vitality and expansive power' (Duckworth 1895, p. 41). Workers were being drawn into the new industries from many sources: from the watch and clock trade for 'metres' (*sic*), from wire-workers and brass-finishers, from electro-platers and india-rubber workers, and 'from every branch of engineering' (ibid., p. 42). These were classically trades of the inner London industrial quarters, in places like Clerkenwell. But, as the new industries drew on their skills, already much of the work was being done beyond the then-London boundary – that is, in what is now outer London.

The interwar years and the 'new industries'

Forty years later, the *New survey of London life and labour* could confirm that this process of growth and movement had continued. In electrical engineering: 'These are the trades which have expanded most rapidly since the war, and which, since they are light trades, have found in the outlying parts of London a suitable field for growth' (Llewellyn Smith 1931, p. 128). London was 'of outstanding importance' in 'telegraphic and telephonic apparatus (including automatic telephone machinery), incandescent lamps, cables and flex, and electrical merchandise such as stoves, lamps, electric irons and kettles, wireless sets and components, and so forth' (ibid.). Generally, the newer and therefore the better equipped firms tend to be found on the outskirts of London. The advantages to be found upon the fringe of the metropolis, at places such as Hayes, Southall and Ponders End, are such as to tempt firms outwards unless special circumstances prevail to keep them in the centre of London. Migrations even farther afield to places like Welwyn Garden City are not unknown. (ibid., pp. 128–9)

D.H. Smith's painstaking account of the industry of north-west London, published in 1933, powerfully confirms the conclusions of the second *Survey*. He covered a sector stretching from the River Lea in the east to the Thames in the south; already, this sector had as many as 140 000 industrial jobs, of which over half (75 000) were in West Middlesex, that is, between the present M1 and the Thames (Smith 1933, p. 173). Of these, some 8000 were in the electrical equipment industry (ibid., p. 105). Out of some 627 factories in the entire sector that Smith was able to analyse, precisely half had migrated there: 313, of which 243 had come from inner London, 43 from abroad (20 from the United States) and 27 from other parts of Britain. The other half, 314 factories, were newly established – no fewer than 232 of them in the decade immediately preceding Smith's study (ibid., p. 171). It is a reasonable supposition that the great bulk of these were started by local entrepreneurs, and therefore represent an indirect form of outward movement.

Not all these firms and jobs were high tech, whether in terms of the technology of that time or in those of today. But Smith was quite specific that many were:

> We are living in a time which might be described as a 'synthetic' period, in which so many of the products which enter into our daily lives are the result of chemical or scientific research . . . Nearness to the consumer seems to be the dominating control over the location of factories of this type. (ibid., p. 105)

And, for the future:

> The forces of scientific progress resulting in increased specialisation will tend to attract particular types of industry towards London, e.g., special machinery, instruments, light engineering, equipment, foodstuffs. (ibid., p. 178)

Some of the industries in these categories would be of particular importance to the subsequent growth of high tech functions. The Gramophone Company at Hayes employed as many as 6000 people (ibid., p. 100), having been established at Maiden Lane in the West End in 1898 and moved out as early as 1907 (Chew 1981, pp. 45, 50). There were other large factories in electrical manufacture: British Thomson Houston at Willesden, General Electric at Hendon, Ediswan at Enfield (Smith 1933, p. 137). Seven out of the eight manufacturers of radio valves had plants in London (ibid., p. 140).

For the middle and late 1930s – a period that is significant because it includes the beginnings of rearmament – a unique series of Board of Trade surveys allows us to extend Smith's analysis both in time and in space. In Figure 7.2, the results are shown for a more restricted definition of contemporary high tech, which includes radio and television technologies, scientific instruments and components, and aircraft. The pattern overwhelmingly confirms the impression of local deconcentration within Greater London. Both new firms – the great majority – and relocations tend to base themselves within the north-west sector of London. There are very few signs of activity in the outer counties, and of these the great majority consist of aircraft firms in the Southampton area. Immediately outside the Greater London boundary, the Slough Trading Estate – where Smith in 1932 already found 12 000 workers in 146 firms, including 46 overseas companies – had attracted one or two high tech enterprises; that was about the sum total.

Nevertheless, contemporary observers seem to have agreed that developments like Slough represented the wave of the future. Its history was rather remarkable: built in the last year of World War I as a lorry dump for the Western Front, it had been located to meet War Office stipulations that it must be near both London and the south coast; be west of London to minimize air raid danger; and be on level

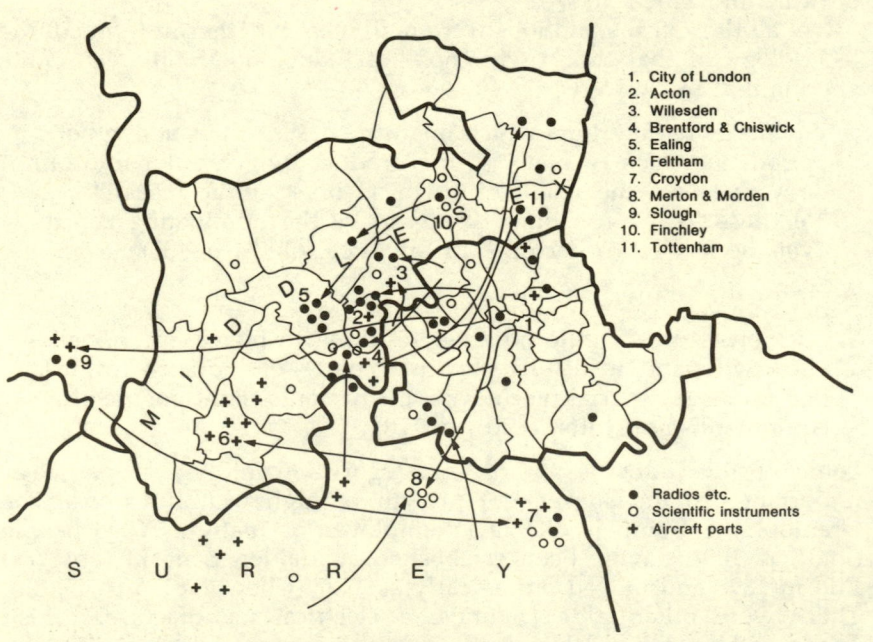

Figure 7.2 High technology industries: new factories and movements, 1933–8.
Source: Board of Trade surveys.

ground with a gravel soil. In a hasty wartime survey the number of possible sites was limited, and Slough came closest. After the war the Slough Trading Company bought the site principally for the lorries; only afterwards did it realize the potential for a trading estate (Hall 1962, p. 167). Noel Mobbs, its developer, had commended it to the Barlow Commission in his evidence of 1938:

> Modern facilities are such that close proximity to the centre of a large town is unnecessary and the principle of satellite towns separated from a centre by a suitable area of open space is quite practical . . . It will require no urge from the Commission to foster this trek. (GB RC Distribution 1938, p. 337)

The effect of World War II

World War II began to confirm this on a small scale. Michael Fogarty, in a survey right at the war's end, stated that:

> In Hertfordshire, the parts of Berkshire and Buckinghamshire within easy reach of London, and around Oxford there has been during the war a considerable influx both of industries and evacuated offices . . . There has been a good deal of new industrial building in the parts of the Home Counties nearest to London, or in districts such as the Kennet Valley between Reading and Newbury, and in most cases the new works are of types which are likely to be of permanent value. (Fogarty 1945, pp. 406–7)

But, he emphasized, this had occurred 'without any dramatic change such has occurred at Slough or Oxford' (ibid., p. 395). Conversely, both in Slough and in west Middlesex, there had been big industrial expansion during the war in engineering, electrical apparatus and related industries. The fact that these developments

> have to a great extent followed pre-war trends is some evidence of their possible permanence . . . In areas such as Hayes, Uxbridge, or . . . Slough, informed observers believe that a good deal of the industrial development which has taken place during the war will survive afterwards. (ibid., p. 439)

The western sector in the 1950s

The 'informed observers' were right: B. A. Bates's study of west London industry, completed in 1952, showed a steady growth since the early 1930s. In Hayes, for instance, electrical engineering employment had increased by over 50% since 1932, from 4910 to 7640 (Bates 1954, p. 91). By that time, in the whole of west London, the incipient electronics sector – wireless, gramophones, records – employed 12 890 workers, in 74 firms. There were 11 firms with more than 200 workers. Apart from EMI at Hayes – by far the biggest – they included Ultra, Dubilier and Pilot in Acton, Hale at Ealing, and Bush at Chiswick. But about half the employment was in small firms with 15 or fewer workers, mainly small parts contractors or radio specialists (ibid., p. 43).

Martin's equally detailed analysis shows clearly that in 1954 the western sector was one of the main areas of concentration of the electrical engineering industries (Martin 1966, p. 99). One principal division was the manufacture of electrical parts for vehicles; another, the manufacture of electrical instruments and of electrical components of all kinds. One manufacturer of industrial valves, with plants in Hammersmith and Wembley, was responsible for half the total British production; its output was bought by firms making hearing aids, computers, telecommunications equipment and laboratory equipment, which were fields where demand was too small to justify the production by firms of their own valves; Martin noticed that it was in such technically advanced or custom-built components that industrial linkages were often confined within south-east England (ibid., p. 102). Medium-scale radio manufacture was also concentrated in London, and 'the re-adaptation of London radio factories and firms is a good illustration of continual shift to newer and more technically advanced or less standardised products' (ibid., p. 101).

In particular, Martin emphasized: 'The field of research and experimental manufacture in electrical engineering is one where London is supreme' (ibid., p. 102). Especially, 'Outer south-west London has emerged in recent years as an incubator area for new firms of extremely rapid growth in advanced electronic industries, a type of role previously more the preserve of central London' (ibid., p. 102): Solartron had migrated from Kingston to Farnborough, Racal from Isleworth to Bracknell.

By that time, the radio industry had already experienced an extraordinary growth. Nationally, employment had increased nearly six times in 20 years: from 13 580 workers in 1930, to 67 100 in 1948 and 90 000 in 1951. As it had grown, so it had decentralized. In 1935, 88.5% of the labour force was in Greater London; by 1951, only 56% was in Greater London but another 17.4% was in the (then) eastern region (Hague & Dunning 1955, p. 203). Firms that had gone further afield reported that they would have preferred to stay in London or build branches close by; many said that they dared not move their main operation far from London, because of the need to recruit skilled labour and to retain close contact with the centre of the industry (ibid., p. 204).

The outward movement of industry

Such reasons remained cogent, even as large numbers of firms did move out of London. When David Keeble made his major study of decentralization from London in the mid-1960s, he found that from an area of north-west London between the M1 and the Thames (and extending roughly two miles (3.2 km) outward from the North Circular Road), 266 factories had moved between 1940 and 1964, taking with them 71 000 jobs: equivalent to 32% of all factories, and 30% of all jobs, in that zone in 1960 (Keeble 1968, pp. 2–3). The outward movement was dominated by factories in metal fabricating and mechanical engineering (45%) and electrical engineering (17%) (ibid., p. 13). The overwhelmingly most important reason given – quoted by 43% as a main reason and by 26% as a subsidiary reason – was shortage of space; shortage of labour and government controls were the next most often mentioned (ibid., p. 5).

Firms moved, then, because they needed space; and, Keeble concludes, this was because they were expanding. Many were involved in the rearmament drive after 1950, or had a good export record; so, even in periods of tight location controls, they found it relatively easy to move. What is most significant is the locations they chose. Nearly 60% of all the moves went to places within 100 miles of London, and of these two-thirds (40% of the total) went to a fairly narrow band between 15 and 40 miles from London: that is, to the Outer Metropolitan Area. And, Keeble notes: 'within this belt, by far the greatest numbers are to be found to the north-west of London, in a broad segment enclosed by lines joining central London to Salisbury in the south and Peterborough in the north' (ibid., p. 26). Among these short-distance movers, the most frequently quoted location factors were proximity to London, availability of a modern factory, proximity to the homes of the directors, and the need to retain the existing labour force (ibid., p. 36).

Keeble thus observed that, in effect, the same forces that drove London industry in the interwar period continued to drive it in the postwar era; the only significant feature was that the planning system now concentrated the outward movement in a relatively few nucleated settlements, the most important of which were Hemel Hempstead, Maidenhead, Slough, Watford and High Wycombe (ibid., p. 46): a feature that, as already seen, Noel Mobbs had accurately predicted in 1938.

Thus, Keeble concluded:

> The fact that North-West London since 1945 has in turn been exporting manufacturing firms still farther to the north-west of the metropolis establishes conclusively the importance of radial movement in the geographical spread of London's industry over the last sixty years. (ibid., p. 46)

If we assume that high tech industries followed the general trend –
and may indeed have done so in intensified fashion – then the postwar
development of the Western Crescent is no more than the interwar
development of the North-West London Sector, writ large. No more
complex explanation is necessary.

The unexplained anomaly

There is, however, one anomalous feature. This is the fact that the key
interwar industrial sector – radio and radio components – was by no
means particularly biased towards the western end of the North-West
London Sector. General Electric's research was in Wembley; Cossor
were in Highbury and Tottenham; Pye in Cambridge; Murphy in
Welwyn Garden City; E. K. Cole in Southend. Seven out of ten valve
makers were in London, but they were widely distributed, from
Ponders End (Edison Swan) to Balham (Mullard) (*Blue Book* 1938,
passim). Only the Gramophone Company (EMI), with their huge
factory at Hayes, gave a distinctly western weight to the distribution.
Smith's analysis of 1933 shows that electrical engineering and
gramophones employed 5350 workers in Wembley, Acton and
Willesden and 6300 in Hayes and Southall, but as many as 6750 in the
Lea Valley (Smith 1933, pp. 51, 94, 99).

On that basis, the postwar electronics industry might just as well
have developed in Welwyn and Stevenage as in Bracknell or Newbury.
It is true that – as Chapter 4 has shown – the Crescent has its
secondary focus at its northern end, in Hertfordshire and especially in
mid-Hertfordshire. This focus, based mainly on aerospace in Stevenage
and Hatfield, may owe something to that earlier north London
concentration; more research would be needed on that point. But by
the 1960s, the primary focus was clearly in the West and South-West,
and for that concentration, there appears to be no clear explanation in
the interwar pattern of industry. If there is an explanation for the
phenomenon of the Western Corridor as distinct from the Western
Crescent, therefore, it must lie elsewhere.

8

The role of the government research establishments

Chapter 7 ended with a mystery: outward movement of London firms provided an explanation for the growth of the Western Crescent, but not for the concentration of high tech industry at its western and southwestern extremities. It is almost as if lack of space created a negative electric current, causing firms to move like iron filings in an electric field – but that some powerful magnet then caused them to coalesce. That magnet, almost certainly, was provided by the government research establishments (GREs), and particularly that group of GREs working for the Ministry of Defence (defence research establishments – DREs). GREs had already developed west of London by the outbreak of World War II, but their number greatly increased during the war and again in the Cold War period of the 1950s, when they began to develop close contractual relationships with high technology industries – especially in electronics production. In this chapter, we therefore look at the origins and the growth of the GREs; in Chapter 9, we shall go on to analyse their relationships with industry.

The geography of the GREs

Table 8.1 lists the 11 DREs after the reorganization and consolidation of the mid-1970s; Table 8.2 gives the more detailed list of the DREs before reorganization, together with a small sample of civil research establishments whose work related to defence, as they existed in the late

Table 8.1 The defence research establishments, 1984.

1 Admiralty Marine Technology Establishment, Teddington, Middlesex	Structural design of ships and submarines, hydrodynamics, vulnerability to above and below surface weapon effects. Civil work in various areas. (Also at Alverstoke, Hants; Dunfermline; Glen Fruin, Helensburgh, Dunbartonshire; Haslar, Gosport; Holton Heath, Poole, Dorset; HM Naval Base, Portsmouth.)
2 Admiralty Underwater Weapons Establishment, Portland, Dorset	Undersea warfare activities, e.g. torpedo research, mine countermeasures, sonar, signal processing etc. (Also at Helston, Cornwall.)
3 Admiralty Surface Weapons Establishment, Portsdown, Portsmouth, Hampshire	Guided weapons, guns, surveillance and tracking systems, communications and navigation. Civil work. (Also Admiralty Compass Observatory, Slough.)
4 Royal Signals and Radar Establishment, Malvern, Worcs	Applied electronics R&D for such systems as airborne and ground radars, guided weapons communication systems and computing. Civil work includes electronics research and maintenance of RF and microwave standards on behalf of the Department of Trade and Industry, and research into air traffic control for the Civil Aviation Authority.
5 Atomic Weapons Research Establishment, Aldermaston, Berks	Mathematics and applied physics, material sciences, explosives etc. Primary concern with the defence nuclear programme, but also involved in other defence and civil research.
6 Chemical Defence Establishment, Porton Down, Salisbury, Wiltshire	Research into, and evaluation of, the threat from use of chemical or biological agents. Research into protection from hostile environments.
7 Military Vehicles and Engineering Establishments, Chertsey, Surrey	Combat and logistic vehicles, engineer equipment and military bridges.
8 Propellants Explosives and Rocket Motor Establishment, Westcott, Bucks	Proposed formulation and performance of motor materials, design and testing. Gun propellants, primary explosives etc. for both defence and civil application. (Also at Waltham Abbey, Essex.)
9 Royal Armament Research and Development Establishment, Fort Halstead, Sevenoaks, Kent	Guns, mortars, mines, warheads, fuses, anti-tank guided weapons, surveillance systems and general land warfare matters.

Table 8.1 *continued*

10 Aeroplane and Armament Experimental Establishment, Boscombe Down, Salisbury, Wiltshire	Testing of military aircraft and associated weapon systems, including arctic and tropical trials. Empire Test Pilots' School.
11 Royal Aircraft Establishment (a) Farnborough, Hants	Military and civil aerospace activities (other than radars), e.g. aerodynamics, avionics, communications, materials and structures, aircraft and systems engineering. (Also at Aberporth, Dyfed; Bedford.)
(b) Pyestock, Farnborough unit	Research and development into, and advice on and testing of, gas turbines for land, sea and air use; also diesel engines and engineering equipment. (Also at West Drayton, Middx.)

Source: GB MOD (n.d.).

1960s – and indeed had existed, with minor changes, since World War II; Figure 8.1 maps their distribution. It is immediately evident that they are extraordinarily concentrated in the Western Crescent, and particularly in places close to the M4 Corridor.

Of the 11 major post-reorganization DREs with significant local purchasing power, eight are in the South-East; two in the South-West; the last is at Malvern in Hereford and Worcestershire. Of the 27 pre-reorganization GREs, Figure 8.1 shows that the great bulk were concentrated in the South-East with a lone establishment in Dunfermline (Law 1983). Following the Strathcona report (GB Steering Group 1980), the DREs are to be concentrated into seven major units, with some current DREs retained only as out-stations. Finally, of the Ministry of Defence's 13 procurement contact points, eight are in London, two in Bath, one in Hampshire and one (clothing and textiles) in Leeds (Table 8.3).

Figure 8.1 cannot show relative size, because systematic data on this point have not been published since World War II; Figure 8.2, however, shows the complement of scientific and experimental personnel on the eve of war. This makes it clear that, at that time, there were five major centres of defence and related R&D: the Whitehall headquarters of the defence ministries; Woolwich Arsenal, traditional centre of research activities for the Army; Teddington, the centre of the National Physical Laboratory and of some Admiralty R&D; Farnborough, the home of the Royal Aircraft Establishment; and the Portsmouth area, the location of most naval R&D. These contained by far the largest R&D establishments of that time: the Royal Aircraft Establishment, the Army Research Department at Woolwich and the National Physical Laboratory at Bushey.

Table 8.2 Government R&D establishments, defence and defence-related, 1968.

Code	Name	Location	Function
1 Navy			
ASWE	Admiralty Surface Weapons Establishment	Portsdown, Hants	Development of ship-borne equipment for guided weapons, including radars, magazine stowage, launchers and handling gear; radars for aircraft direction, early warning and navigation; communications and other electronic warfare equipment; action data automation.
AUWE	Admiralty Underwater Weapons Establishment	Portland, Dorset	Development of torpedoes, their control systems and launching gear; sonar equipment and associated trainers; mine counter-measures and research into problems such as acoustic propagation, reverberation and the use of computer techniques; experimental diving.
ARE	Admiralty Research Establishment	Teddington, Middlesex	Research into underwater acoustics, fluid dynamics including drag and water entry techniques, instrumentation, optics and general supporting mathematical studies; studies of radiological problems.
SERL	Services Electronics Research Laboratory	Baldock, Herts	Microwave electronics, semiconductors, transistors, lasers and gaseous electronics for all three services.
AML	Admiralty Materials Laboratory	Holton Heath, Dorset	Research into aspects of materials and chemical engineering which are of special interest to the RN. Fuel cell research work for all three services.

Table 8.2 *continued*

Code	Name	Location	Function
ACO	Admiralty Compass Observatory	Slough, Bucks	Advanced navigational systems, including inertial navigation, gyro and magnetic compasses. Compass testing for all three services.
AEL	Admiralty Engineering Laboratory	West Drayton, Middlesex	Test, evaluation, and development of internal combustion engines and electrical equipment for HM ships; instrumentation and control systems; noise and vibration problems.
AMEE	Admiralty Marine Engineering Establishment	Haslar, Hants	Test and evaluation of ship-borne machinery other than internal combustion engines, including boiler controls, combustion equipment and auxiliary machinery.
AEW	Admiralty Experiment Works	Haslar, Hants	Hull design and equipment, including model testing. Manoeuvrability and control, seaworthiness and stabilization, dynamic stability of surface ships submarines. Ship trials and analysis; propeller design and noise reduction.
NCRE	Naval Construction Research Establishment	Dunfermline, Fife	Structural design, strength tests and analysis. Explosion and shock phenomena and effects. Noise transmission through structures. Development of structural materials and welding techniques.

continued

Table 8.2 *continued*

Code	Name	Location	Function
2 Army			
RARDE	Royal Armament Research and Development Establishment	Sevenoaks, Kent	Development of weapons and weapon systems for the Army; some weapon and equipment development for all three services, and work for the Home Office on industrial explosives and hazardous substances and investigations arising from the use of explosives for criminal purposes. Work on guided weapons for the Ministry of Technology.
FVRDE	Fighting Vehicles Research and Development Establishment	Chertsey Surrey	Development of armoured vehicles for the Army; also special vehicles for all services including the modification of commercial vehicles for defence applications and the assessment and evaluation of commercial vehicles and passenger cars.
MEEE	Military Engineering Experimental Establishment	Christchurch, Hants	Development of military engineer equipment; may also, on a commercial basis, evaluate commercial construction plant and equipment.
CDEE	Chemical Defence Experimental Establishment	Porton Down, Wilts	Assessment of the hazards of chemical warfare, and development of chemical defence equipment on behalf of the three services and Civil Defence.
MRE	Microbiological Research Establishment	Porton Down, Wilts	Assessment of the risk of biological warfare and the design of means of defence.

Table 8.2 *continued*

Code	Name	Location	Function
APRE	Army Personnel Research Establishment	Farnborough, Hants	All aspects of human factors research for the Army.
SCRDE	Stores and Clothing Research and Development Establishment	Colchester, Essex	Research and development of clothing and general stores for the three services.
PREEs	Proof and Experimental Establishments	Shoeburyness, Essex	Trials required in connection with research, development and production of explosive stores, gun barrels and the like.
OB	Ordnance Board	London	An independent inter-service technical trials and advisory organization, whose costs are borne by Army Votes, for the approval of the safety and effectiveness of weapons and weapon systems in which explosives are used; monthly tests are held in Proof and Experimental Establishments.

continued

Table 8.2 *continued*

Code	Name	Location	Function
3 Avmin			
RAE	Royal Aircraft Establishment	Farnborough, Hants	General responsibilities for R&D on civil and military aircraft and spacecraft; work on specific civil and military projects, both national and collaborative; work (largely of civil application) on air traffic control, traffic analysis, airport equipment, safety and accident investigation; basic research on technology of general application, e.g. development of carbon fibre composites and use of silicon in electronics, blind landing, head-up display, moving map display, slender wing concept for supersonic aircraft.
NGTE	National Gas Turbine Establishment	Pyestock, Hants	Research into the problems of gas turbine engines and related systems, and support of their development by industry, including full-scale engine testing.
A&AEE	Aeroplane and Armament Experimental Establishment	Boscombe Down, Wilts	Proving and acceptance testing of military aircraft and aircraft equipment, including armament, radio and navigation equipment.
ATDU	Aircraft Torpedo Development Unit	Helston, Cornwall	Development of naval air armament.

Table 8.2 *continued*

Code	Name	Location	Function
RRE	Royal Radar Establishment	Malvern, Worcs	Research and development in ground and air-borne radar systems and the electronic and systems aspects of guided weapons R&D; basic work on quantum electronics and solid state physics. Important support to the electronics industry and to the wide range of industries in which the products and techniques of the electronics industry can be applied. Development of much of the equipment for air traffic control; research into radio astronomy.
ERDE	Explosives Research and Development Establishment	Waltham Abbey, Herts	Investigation of non-nuclear military explosives, rocket propellants and initiating devices and non-metallic materials of both organic and ceramic type.
RPE	Rocket Propulsion Establishment	Westcott, Bucks	Design and development of rocket motors for guided missiles and space research vehicles.
SRDE	Signals Research and Development Establishment	Christchurch, Hants	Telecommunications research and development, mainly for defence, including collaborative work with other countries; infra-red viewing devices; provision of intensified images; fuel cell development for specific purposes; thermo-electric generation, radioactive isotype power sources and mobile turbo-alternators.

continued

Table 8.2   *continued*

Code	Name	Location	Function
NPL	National Physical Laboratory	Bushey, Middlesex	Aero Department is substantially engaged on a joint research programme with RAE.

Source: GB SC Science 1969.

Table 8.3   Ministry of Defence contact points, 1983.

General Contracts Queries: London
Sea Systems Equipment: Bath, Portsdown, Hants; Portland, Dorset
Land Systems Equipment: London
Air Systems Equipment: London (two offices)
Research & Development: London (two offices); plus the individual research establishments
Royal Ordnance Factories: London; individual ROFs
General Naval Supplies: Bath
Clothing and Textiles: Leeds
Navy, Army, Air Force Institutes (NAAFIs): London

Source: GB MOD (n.d.).

Since then, without doubt, there have been some important shifts: in particular, the development of the Royal Aircraft Establishment's branch facility at Bedford, the decentralization of the Army's R&D Establishment from Woolwich Arsenal to Fort Halstead at Sevenoaks, the establishment of the Radar Research Establishment at Malvern, and the growth of the Atomic Energy complexes at Harwell and Aldermaston. Despite these, there seems little doubt that the Western Crescent has retained a dominant role.

By 1968, the date of the Select Committee's investigation on which Table 8.2 is based, there were – apart from the Atomic Energy Authority's establishments – 26 distinct defence research establishments under the direction of either the Ministry of Defence or the Ministry of Technology (formerly the Ministry of Supply). Though systematic data on size are lacking, statements show that the largest

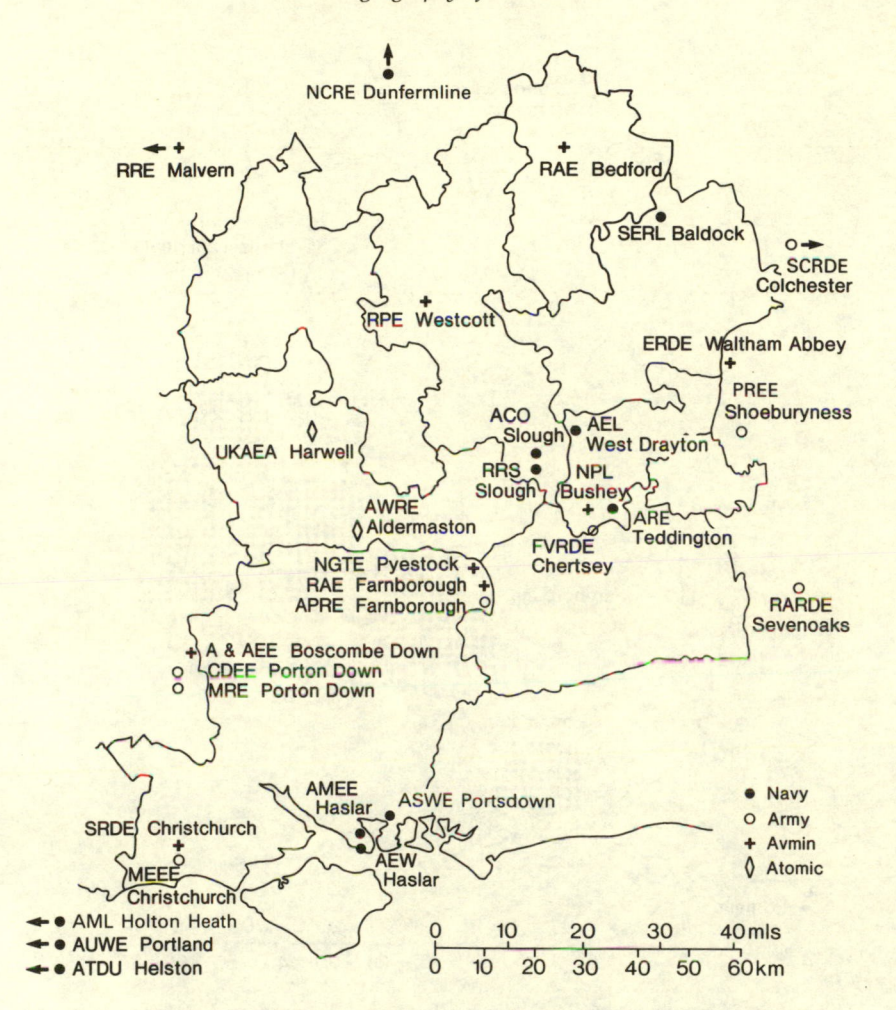

Figure 8.1 Government R&D establishments, defence
and defence-related, 1968.
Source: GB Select Committee on Science and Technology (1969).

were the Royal Aircraft Establishment, with over 1400 professional staff
(GB SC Science 1969, p. xi); the Atomic Energy Research Establishment, with some 1280 (Clarke 1981, p. 7); and the Radar Research
Establishment, with 550 (GB SC Science 1969, p. 288); at the other end of
the scale was the Aircraft Torpedo Development Unit with seven (ibid.,
p. xi). Some few, like the Royal Aircraft Establishment, dated from
before World War I; others from 1914–18 or just after, like the Chemical

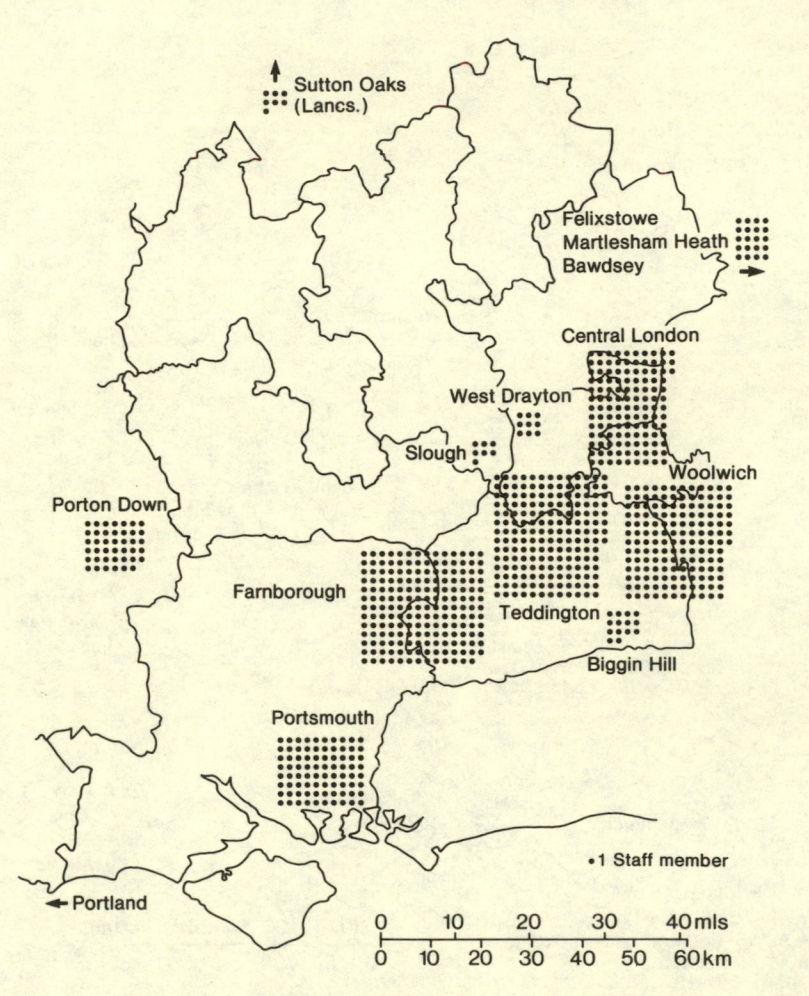

Sutton Oaks
(Lancs.)

Felixstowe
Martlesham Heath
Bawdsey

Central London

West Drayton

Slough

Woolwich

Porton Down

Farnborough

Teddington

Biggin Hill

Portsmouth

•1 Staff member

Portland

0   10   20   30   40 mls
0 10 20 30 40 50 60 km

Figure 8.2   Government R&D establishments, defence and defence-related:
scientific and experimental staff, 1938.
Source: Imperial Calendar.

Defence Establishment at Porton (1916) or the Admiralty Research
Laboratory at Teddington (1920). But most dated from World War II,
when the government found that in order to keep up with the rapid
advances in military technology it could not rely on industrial research
(ibid.).

All the establishments, the Electronic Engineering Association told
the Select Committee, 'have always worked in close collaboration with
the military machine' (ibid., p. 177). But some – for instance the Royal
Aircraft Establishment and the nearby National Gas Turbine Estab-

lishment – also served civil needs (ibid., pp. xi-xii). They all had three main purposes: basic research; the evaluation of work done under contract by industry; and the provision of specialized capital equipment or techniques that private industry lacked (ibid., p. xii). Their total extramural expenditure, at £189 million, was then more than double their in-house expenditure of £71 million (ibid., p. xiii): a clear indication of the spin-off from defence work to industry.

The location of the defence establishments seems to have followed a pattern of incremental growth: they started life as offshoots from established military activities in existing defence locations, and then either grew *in situ* or moved a short distance in search of space. Only occasionally, under the stress of total war, did they undertake major relocations for security reasons. The Admiralty establishments provide a classic case. Traditionally, the Navy had built its own ships in its own dockyards; in World War II, unlike the other two services, it did not give up its supply functions to specialized ministries; after the war, when these ministries amalgamated to form the Ministry of Supply, it continued to remain aloof (GB SC Estimates 1947, p. viii). Thus, beginning with the establishment of the Admiralty Research Laboratory in 1920, it had created its own proliferating network of research and development establishments, a fact that exercised one member of a Parliamentary committee to comment: 'The Admiralty always seems to want to run its own show, and it does make us suspicious that there is a certain amount of "empire building" behind it' (GB SC Estimates 1953, p. 107).

Despite a strong recommendation from a further Estimates Committee in 1956 that there should be immediate examination of the possible merger of naval R&D establishments and those in the rest of the defence machine (GB SC Estimates 1956, p. xxvii), the situation remained unchanged in the late 1960s, when yet another House of Commons Committee resignedly declared that it was a product of history (GB SC Science 1969, pp. 184–5). The result was that the Admiralty establishments formed a solid group in the Portsmouth area, with out-stations a few miles away in Dorset.

In the other services, though traditions were less set, the same principle of inertia applied. This pattern is well illustrated by the histories of some of the most important establishments.

The Royal Aircraft Establishment

The story of the RAE begins at Woolwich, the traditional home of the British Army's armaments, in 1878. In that year, conscious of the possible role of balloons in military reconnaissance, the War Office opened a Balloon Equipment Store under the Royal Engineers there; in 1882 it sanctioned the establishment of a Balloon Factory at the Royal Engineers' base at nearby Chatham; in 1888, after the successful use of

balloons in the Sudan and Bechuanaland campaigns, it established a
Balloon School there. In 1890 the War Office sanctioned the establishment of a regular Balloon Section, and in 1892 a Balloon School and
associated Balloon Factory were set up as part of the Royal Engineers
Gibraltar Barracks complex south of the Aldershot Canal in central
Aldershot, at a spot later renamed Balloon Square (GB RAE 1955, p. 4,
back cover; Walker 1971, pp. 2, 5; Boyd 1975, p. 68; Cole 1980, pp. 134,
339).

Aldershot by that time was the main headquarters and training
centre of the British Army, with accommodation for over 20 000 troops
and a huge area of training grounds stretching west to Fleet, Crundall
and Cookham (Cole 1980, pp. 82, 109). It resulted from a conscious
locational decision in 1853 by Lord Hardinge, then Commander-in-
Chief, who after survey concluded that it was:

> admirably adapted for the assembly of a large military force, from
> the interior moving to and from London by two railways in direct
> communication with Portsmouth, Chatham and Dover; for all
> purposes of strategy it is one of the most important points that
> could be selected, with an ample supply of water at all seasons.
> The tract of land is therefore suited for a permanent camp of
> Instruction in peace and of concentration in War . . . I do not
> believe, that any waste land possessing the great advantages of
> Aldershot from its position can be found in any other of the
> Maritime Counties. (Quoted in Cole 1980, p. 28.)

Acting with unusual alacrity under stimulus of the Crimean War, the
Treasury in January 1854 approved the expenditure of £100 000 for the
purchase of 25 000 acres (10 000 ha) to establish a permanent camp,
which began construction in September of that year and was
completed by 1859 (ibid., pp. 29, 31).

The extent of this purchase proved highly significant. One part of it,
Laffan's Plain to the west of the Farnborough Road, was a giant
ceremonial parade ground named after Major-General Robert Laffan,
the planner of Aldershot (Walker 1971, p. 59). In 1903, after an
investigation that surveyed sites as distant as Weedon in Northamp-
tonshire, a Committee of Inquiry recommended the removal of the
Balloon School and Factory to this site; the impetus was the sudden
expansion of military ballooning in the South African War, which had
made the existing 2.5-acre (1 ha) site hopelessly too small (ibid.,
pp. 41–3, 53). The Committee was particularly impressed by the need
to build a large shed for airship construction , or, as it was officially
known, an Elongated Balloon Erecting House (ibid., pp. 55–6, 60). The
resulting site, to which the School and Factory moved in 1905–6, had
1500 acres (600 ha) of reasonably level open common land, admirably
suited for eventual expansion. As the historian of the RAE has put it:
'Thus it came about that the balloon men of 1904, in seeking a place for
an airship shed, unconsciously laid down plans for an establishment

that has proved invaluable in two world wars and is still one of Britain's greatest assets' (ibid., p. 57).

Soon, the Balloon School was moving from airships to airplanes. S. F. Cody, the illiterate American showman-adventurer who came to play a key role in the RAE's early history, made the first powered flight here in 1908 (Walker 1974, pp. 79–80). A year later, the Army Balloon Factory was given a measure of civilian control, working closely with the newly formed Advisory Committee for Aeronautics of the National Physical Laboratory; the Balloon School remained a Royal Engineers unit until 1911, when it became an RE Batallion and, a year later, the Royal Flying Corps (RFC) – the precursor of the modern RAF; a year after that, the Factory was transferred to the RFC and became the Royal Aircraft Factory (GB RAE 1955, pp. 4–5, back cover).

World War I saw a major expansion of the Factory; but after pressure from private manufacturers, a 1916 judicial inquiry and a 1924 Committee of Inquiry resulted in a critical decision that production of aircraft should be their prerogative, leaving Farnborough exclusively as a research and development centre (ibid., p. 5; Postan *et al*. 1964, p. 438). It was from that point that a close symbiotic relationship began to develop between the Farnborough establishment – now renamed the Royal Aircraft Establishment – and the private manufacturers. During the interwar period, it became 'established practice for aircraft firms to evolve and develop their designs with the accumulated experience and facilities at Farnborough' (GB RAE 1955, p. 10).

Even by World War I, an increasing part of the RAE's work was on instrumentation. After World War II, the work grew and changed with the coming of the atomic warhead, guided weapons and new forms of propulsion. The Establishment outgrew its 1100-acre (440 ha) site and began to develop a new out-station at Bedford (ibid., pp. 14, 16). But in addition, a host of other establishments developed during and immediately after the war for specialized purposes, with some degree of help from RAE: the National Gas Turbine Establishment at Pyestock, next door on the Farnborough site; the Rocket Propulsion establishment at Westcott near Aylesbury, Buckinghamshire; the Aeroplane and Armament Experimental Establishment at Boscombe Down near Salisbury, Wiltshire; and the Royal Radar Establishment at Malvern, Worcestershire (Lighthill 1965, p. 29).

Except for Malvern, all these sites are in the Western Crescent. It seems clear that they resulted from ad hoc decisions taken in World War II. Boscombe Down is part of a complex of research establishments on Salisbury Plain north-west of Salisbury, the oldest of which (the Chemical Defence Research Establishment) was established at nearby Porton as early as 1916 to study the effects of poison gas in World War I. Between 1897 and 1902, presumably in response to the need for additional training grounds during the South African War, the War Office had bought no fewer than 43 500 acres (17 400 ha) of Salisbury Plain including the chief estates of Tidworth, Burford, Durrington and

Ludgershall and the downland around Larkhill, which now house a complex of barracks, training and testing grounds and research establishments (Whitlock 1976, p. 117). Logically, because of its size and relative isolation from large population centres, the area came to house those research establishments that involved the most dangerous or noisy testing procedures.

The National Physical Laboratory

At almost the same time as the Balloon School was making its moves first to Aldershot and then to Farnborough, another equally momentous development was occurring slightly closer to London. During the 1880s and 1890s there was intense concern at the rapid competitive advance of Germany in the newer high technology industries of the day, especially in chemicals and in electrical engineering, which – it was argued – had been aided by the establishment of the Physikalisch-Technische Reichsanstalt at Charlottenburg, Berlin, as a governmental research and testing laboratory (GB DSIR 1951, pp. 16–17; Sutherland 1965, p. 6; Pyatt 1983, pp. 15–16). A committee of inquiry under Lord Rayleigh, in 1897–8, had heard from a representative of British industry that 'we can take it for granted that the advancement of science means in the long run the advancement of industry' (quoted in Pyatt 1983, p. 19).

The argument proved decisive, and the National Physical Laboratory – consciously modelled on the Reichsanstalt – came into existence on 1 January 1900 under the direction of Richard Glazebrook, who himself put on record shortly afterwards that:

> The first aim of the Laboratory is to assist in promoting a union which is certainly necessary if England is to retain her supremacy in trade and in manufacture, to make the forces of science available to the nation . . . and to point out plainly the plan that must be followed unless we are prepared to see our rivals take our place. (Quoted in ibid., p. 31.)

Originally, it was planned that the new laboratory should be in the Old Deer Park, Richmond, close to the old Kew Observatory. But a well-organized local protest caused its last-minute removal to Bushy House in Teddington, a Royal grace-and-favour residence in a corner of Hampton Court park, built in 1663 by a courtier of Charles II (ibid., pp. 3, 22; GB DSIR 1951, pp. 18–19, 22). The conversion was complete, and NPL moved into Bushy House, in 1902.

Soon, the Laboratory outgrew it. The volume of work, particularly in the electrical field, was so great that the first major extension, the electrotechnics building, opened in 1906 (ibid., p. 46). Thence the expansion was steady until World War I, when – faced with a crisis when Britain discovered that it lacked German scientific products – an explosion of war-related work occurred, causing a growth in the staff from 187 at the outbreak of war to 532 at the end of it (ibid., p. 62). In the middle of the war, a new Electrical Department was 'carved off' from the Physics Department (ibid., p. 91). In turn, transferred to the new Department of Science and Industrial Research (DSIR) after 1918, the Laboratory began rapidly to expand its work on radio, which was boosted in 1920 by the establishment of the Radio Research Board, a joint military–civil research body to assist the co-ordination of radio research by the Armed Forces and the Post Office (GB DSIR 1930, p. 1). At that point, an out-station was established for this purpose at Ditton Park, Slough:

> Owing to its situation and the proximity of other electrical work, however, the National Physical Laboratory is unsuitable for the conduct of radio research work requiring measurement in the field or on isolated sites. For such observations facilities were provided by the Lords Commissioners of the Admiralty on land adjoining their Compass Observatory at Ditton Park, near Slough. A flat and open site of a considerable area was here available, on which could be erected wooden huts so isolated as to enable several experiments to be carried out simultaneously on the propagation of waves and directional wireless. (ibid., p. 2)

Ditton Park, like Bushy House, was a surplus royal residence – the place where Mary I, the Tudor Queen, had spent part of her childhood (Watson-Watt 1957, p. 83). It was presumably chosen for its relative nearness to Teddington, since some of the staff were involved in regular trips between the two (ibid., p. 70; GB DSIR 1930, p. 3). Its subsequent significance, however, was almost the result of an accident.

The Royal Radar Establishment

We already noticed that during World War I the then Royal Aircraft Factory moved rapidly into the field of aircraft instrumentation. Thus, in 1915, the young Robert Watson-Watt was hired from academia to undertake research on radio warnings of thunderstorms. A year later, he

> took possession of the 50 foot by 16 foot hut which a benevolent Meteorological Office had provided as married quarters, to enable me to devote more of my 'leisure' to the radio equipment under whose antennas the new hut now stood. (Watson-Watt 1957, p. 45)

This was Smallshot Bottom, in Aldershot's North Camp. In 1920, with a new management at RAE proving hostile to radio research, he was fortunate in that the newly formed Radio Research Board, associated with the also new Department of Scientific and Industrial Research, funded him to do research on atmospherics: 'Smallshot Bottom proving too small to hold us, we spread a little way further down the Basingstoke Canal, to put on Smallshot Hill our main buildings' (ibid., p. 55).

Officially, therefore, Watson-Watt was now a civil scientist. But, as he makes clear, he benefited immeasurably from close military contacts. The first chairman of the newly formed Radio Research Board was Admiral of the Fleet Sir Henry Jackson, the pioneer of naval radio telegraphy; antennae were built for him by Colour-Sergeant Joe Carey, on the advice of the Army representative on the Board (ibid., p. 55).

At Aldershot, Watson-Watt and his research were still occupying the Radio Station of the Air Ministry's Meteorological Office, which had been taken over by the new Radio Research Board, but which was still on War Department land. In 1924 the War Department announced that they wished to reoccupy it, and Watson-Watt was transferred to the Radio Research Station at Slough, where the radio research out-station of the National Physical Laboratory had recently been established; in 1927 he became Superintendent of the entire Station, and in 1933 Director of the NPL's newly created Department of Radio Research at Slough (ibid., pp. 70, 72).

The move proved fateful. For here, in 1935, Watson-Watt and his team made the experiments that proved the feasibility of radar (ibid. p. 83). The next year, Watson-Watt joined the RAF as Superintendent of Radio Detection, in a newly established headquarters at Bawdsey Manor outside Felixstowe in Suffolk – a site chosen because of its nearness to a disused Naval radio direction-finding station on Orfordness, which the Radio Research Board had taken over in the 1920s (ibid., pp. 140, 145; GB DSIR 1930, p. 3). Here, at the Bawdsey Research Station, the first tests of radar were made, and work started to develop a chain of radar stations on Britain's vulnerable east and south coasts. The Bawdsey location was determined by the vital need to set up an operational defensive radar chain along the coast before the outbreak of war, coupled with the existence of radio facilities there.

But, once war had broken out, the site was apparently thought too vulnerable to attack and the establishment was moved to Swanage, Dorset. Finally, after a major air raid on the French coast in July 1942, it was determined that the risk of retaliation on Swanage was too great, and the whole operation was moved to Malvern (Jones 1978, p. 247). Critical, here, was probably an Air Council designation of 1934, which stated that relatively safe areas were those lying north and west of a line from Weston super Mare through Stow on the Wold and Stratford upon Avon to Stockport (Hornby 1958, p. 286). Malvern, as its then director admitted in 1953, was:

> not an ideal site. Technically, you have a mountain behind you, but the capital cost involved in moving it is out of the question; and it is not only a matter of capital cost involved but also the

question of moving staff and their families. It is quite out of the question. (GB SC Estimates 1953, p. 103)

So the establishment that, in the words of the questioner, 'has always suffered from being pushed about and moved from place to place', seemed destined to stay put; which, in the intervening years, it has. The biggest change that has occurred is yet another change in name: the Royal Signal and Radar Establishment, having spent a critical part of its early life in the heart of the M4 Corridor – the motorway was constructed right beside Ditton Park – is now by far the biggest GRE located right outside it. It is the exception that proves the rule.

There is a certain irony in this. For the RAE, having said goodbye to Watson-Watt in 1924, had immediately set up a Radio Department of its own; and during World War II it was in repeated conflict with Watson-Watt's station about its boundaries of responsibility. Ever since 1924, it had been supposed to concentrate on development while the NPL and the Radio Research Station did the basic work; and this was confirmed in the wartime years. RAE scientists clearly resented their relegation to what they thought was the less interesting work (Postan *et al.* 1964, pp. 452–5). But, in terms of the development of contacts with manufacturers and the consequent industrial spin-off, RAE had probably kept the critical part of the work in the Crescent.

The Atomic Energy Research Establishment

The Atomic Energy Research Establishment at Harwell, near Abingdon, resulted from critical decisions taken at the end of World War II. During the war, in July 1941, a committee of government scientists had concluded that it would be possible to develop atomic energy for military purposes by using the gaseous diffusion method for separation of the 235 isotope (Gowing 1964, pp. 76–7). F.A. Lindemann, later Lord Cherwell, Churchill's scientific adviser – who, before entering government, had directed the Clarendon Laboratory at the University of Oxford – pushed the government to develop the work through the DSIR's Department of Tube Alloys, a cover name (ibid., p. 109; GB Ministry of Supply 1954, p. 3; Smith 1961, pp. 295–7). Soon after the war's end, on 29 October 1945, Mr Attlee announced to the House of Commons that the government had decided to create a research and development establishment, to cover all aspects of atomic energy, on an RAF airfield at Harwell in what was then Berkshire (GB Ministry of Supply 1952, p. 10; GB Ministry of Supply 1954, p. 5).

The search for the site of the new establishment had begun towards the end of 1944. It early became clear that given the shortage of facilities and materials at the end of the war, the only practicable site would be a disused permanent airfield complete with engineering workshops, roads, housing, water supply and, above all, hangars to house the large nuclear machines. Such an airfield must be near Oxford or Cambridge, the site of the two major physics laboratories. As between these, Cambridge was the clear favourite; John Cockcroft, later (summer 1945) appointed director (Gowing 1964, p. 333), came from the Cavendish Laboratory in Cambridge, which had a far greater prewar reputation in atomic physics than Oxford's Clarendon Laboratory (ibid., ch. 1 *passim*); so did several other obvious candidates for senior positions.

But in early 1945, the Air Ministry were reluctant to release East Anglian airfields because of their strategic importance, and those under offer – Deben and Duxford – were ruled out, the first because of its high water table, the second because of its poor water supply and remoteness. The inspection team then visited Benson in Oxfordshire, which they favoured until the Station Commander pointed out that the area was quite populous and that Harwell, nearby, had open country around it. Harwell proved good for housing in Abingdon and Wantage, while Oxford and Abingdon offered good schools, important in attracting senior staff; there was an efficient train service to London and Birmingham, and Oxford University had maintained atomic research throughout the war – a crucial consideration in attracting top-quality scientists away from their academic laboratories. So, despite an icy official reception, Cockcroft and his colleagues took the effective decision in Harwell's favour one evening in the nearby 'Horse and Jockey' (GB Ministry of Supply 1952, p. 10; Clarke 1981, pp. 5–6).

Soon, under Cockcroft's direction, Harwell began to evolve into one of the largest research establishments in the country; by the mid-1950s it had 94 buildings and some 1265 scientific staff (GB UKAEA 1958, p. 6; Vick 1965, p. 55; Clarke 1981, p. 7). By this time, spurred by the need to increase the programme in the wake of the Suez oil crisis of 1956, it became clear that the whole complex was too big to remain under a single direction. The newly formed Atomic Energy Authority, which was said to be the single-handed creation of Lord Cherwell (Smith 1961, p. 315), had taken Harwell over in 1954; now, it began to hive off some of its activities into separate newly formed establishments. But, for reasons of convenience, most were set up nearby: high-energy nuclear physics at the adjacent Rutherford High Energy Laboratory, the physics of nuclear fusion to Culham eight miles (13 km) away, the industrial use of isotopes at Wantage, and – the only longer-distance move – reactor physics to Winfrith in Dorset (GB AERE 1980, p. 5; Clarke 1981, p. 7). Two smaller research units, concerned with the effects of radiation and with radiation protection, also located in the area; and there are two private laboratories (Clarke 1981, p. 7). Thus, just as at Farnborough, a research complex began to form. And in turn, the decision in 1949 to consolidate all work on atomic weapons, including related electronic instrumentation, at the disused airfield at Aldermaston, outside Reading (GB UKAEA 1959, p. 6), may well have been influenced by the nearby location of Harwell.

This point is crucial, because it implies a powerful inertia in the location process: once a critical triggering decision is taken, whether carefully or capriciously, subsequent growth will take place at or around that point. A postscript illustrates the principle. In November 1985 the government announced the creation of a British National Space Centre, to develop research in space travel and exploration, to be based initially at the Department of Trade and Industry's London headquarters. The main work underpinning the new strategy, it was announced, would be divided between the Royal Aircraft Establishment at Farnborough and the Rutherford Laboratory at Harwell.

Summing up

The history of the GREs is thus a curious blend of purposive location and accident. Aldershot, which proved so crucial to the development of the RAE and its offspring, was the result of a mid-Victorian rational location search. The Royal Signals Establishment was moved away from the Crescent because of its supposed wartime vulnerability – though other, equally important, GREs were not. The move of the Balloon School to Farnborough after the Boer War, and the establishment of Harwell at the end of World War II, were both the result of systematic locational surveys. Yet there was also an element of accident and of caprice: the Navy's historic presence at Portsmouth, which locked its research locations into that area; the suddenness of the British government's commitment to Aldershot, taken in the press of the Crimean War; the presence of surplus royal residences at Bushy and Ditton; the detachment of Watson-Watt from Farnborough and his move to Slough.

Underlying their location, though, there was a deeper logic. London was the centre of the military and the scientific establishments, both increasingly powerful in the late 19th and early 20th centuries. The need to defend London from potential foreign attack – whether invasion from the Channel, or air attack – was paramount, and it suggested a concentration of defence facilities west and south-west of the capital. Scientific research at the turn of the century was overwhelmingly concentrated in the universities of London, Oxford and Cambridge. Even the availability of surplus royal residences had a certain logic in the hunting pastimes of Tudor and Stuart monarchs. All pointed to locations in the zone west and south-west of London: precisely, the sector between Portsmouth to the south-west and Oxford to the north-west. The M4, once built, bisected this sector.

Until World War II the GREs remained small and not particularly influential, save perhaps in the area of aircraft procurement. But after 1945, with the rapid development of a military–industrial complex in Britain, they quickly came to form the nodes of an intricate web of defence procurement, in which close and intimate contacts between research establishments and high tech contractors became the everyday rule. In the following chapter we shall trace the evolution of these linkages and then analyse their present patterns.

9

The defence contractors

By 1945 the government research establishments were part of the landscape of southern England. But their real significance for the Western Crescent came after this, with the growth of Britain's military–industrial complex in the 1950s. In this chapter we describe and analyse this growth and its significance for the development of the Crescent and the Corridor. First, directly continuing the story of Chapter 8, we chronicle the historical development of GRE–industry relationships during the 1940s and 1950s. Secondly, we outline the pattern of defence support to industry in recent years, with special stress on its clear regional bias towards the south – a bias which has totally contradicted the aims of regional policy. Thirdly, we seek to explain this bias in terms of the peculiar characteristics of the defence equipment procurement process. We illustrate this process from a variety of evidence, including our own survey interviews. And finally, we seek to draw some general conclusions.

1940–1955: The origins of the military-industrial complex

If we delve into history, we can trace the first small beginnings of GRE–industry relationships back to World War I. Watson-Watt, inventor of radar, recalls in his autobiography that as early as 1916, in order to get a cathode-ray tube of the necessary quality, the Royal Aircraft Establishment had to go to A.C. Cossor, then primarily makers of laboratory instruments, with whom Watson-Watt was later to have close contact in the development of radar (Watson-Watt 1957, p. 59).

But this was somewhat premature. The first major expansion of electronics manufacture came not in World War I but in World War II – and in particular after 1941, when the development of centimetric radar, itself made possible by the invention of the cavity magnetron valve, caused a huge surge in military demand. This was met by the conversion of the 1930s radio industry. By 1944 there were 120 000 workers in the radio industry; the industry worked all out to expand production from 12 million valves in 1940 to 35 million in 1944, an additional 17 million coming from the United States (Postan 1952, pp. 365, 369). The need was met simply by enlargements to existing floorspace, thereby reinforcing the prewar location of the industry in and around London; A. C. Cossor and E. K. Cole both expanded their floorspace four times between 1934 and 1941 alone (ibid., p. 369; Hornby 1958, pp. 276–7).

In the production of the radar equipment itself, EMI and GEC – both London firms – played a critical role (Postan 1952, p. 367). Though in 1943 the industry was advised to go to places where labour supplies were easier – Aberdeen, Edinburgh, Llanelli and Wigan were mentioned – it was too late, and the pressures too great, to do much about it (ibid., p. 369). Similarly, in aircraft instrumentation generally, specialized electrical engineering and scientific instrument firms grew very quickly to meet the demand (Hornby 1958, p. 278). The result was a tremendous boost to the electrical industry, not merely in direct employment but in terms of the research contribution to subsequent development in radio, radar and electronics generally; the component industry alone doubled in the decade from 1937 to 1947, from 12 000 to 25 000 workers (*Electrical and Radio Trading* 1947, p. 87).

In all this, however, the industry was very largely confined to pure production. Generally, considered as a whole, it lacked good research facilities; the development work had to be done in the government establishments (Postan *et al.* 1964, p. 429). The main exceptions were GEC (for valves), British Thomson-Houston, Metropolitan-Vickers and Standard Telephone and Cable, plus Electrical and Musical Instruments which was then the research arm of the Gramophone Company; some middle-sized companies, like Cossor, Pye, Murphy, and Cole, also had small but skilled research teams (ibid.) So it was only in special cases, as in EMI's work on centimetric radar, that industry made a substantial contribution to development (ibid., p. 430).

From the war's end to 1951, the radio industry was mainly engaged in readapting to meet the backlog of civilian demand in the face of constant materials shortages. Then, in January 1951, against the background of the Korean War, a Defence White Paper announced a doubling of production of combat aircraft and a fourfold increase in defence spending generally; although this expansion was later scaled back because of economic crisis, nevertheless defence spending doubled in real terms between 1950–1 and 1953–4 (GB Parliament 1951, p. 5; James 1976, pp. 41, 51). This in turn involved rapid expansion of aircraft, radar and radar equipment at a time when home demand was also growing due to the rapid expansion of domestic television (GB Parliament 1951, pp. 6–7).

Even at the start of this buildup, some sections of the radio industry were heavily engaged in defence work. The managing director of Murphy Radio estimated that for the whole industry armaments accounted for £6 million out of a total of £75 million sales, or about 8%; but Marconi said that all their sales were to official bodies and governments, 60% going to the British government (GB SC Estimates 1951, pp. 90–1). Much of the exports – estimated for the whole industry at £18 million or about one-quarter of sales – had been built up since the war with active Admiralty or Ministry of Supply encouragement. The man from Marconi said:

> they fit into this Defence programme, in regard to radar particularly; and they are virtually British Government designs of the last war with a certain amount of re-engineering, and I suppose it is in the national interest that they should be exported. (GB SC Estimates 1951, p. 95)

By 1953, according to the Controller of Guided Weapons and Electronics at the Ministry of Supply, more than one-third of total electronic output – £47 million out of £130–140 million – represented rearmament. That compared with £7 million out of a total of some £95 million, a mere 7%, in 1949–50 (GB SC Estimates 1953, p. 92). The curve of growth would flatten, but would continue:

> because the proportion of electronics within defence is growing very fast indeed . . . if you take what it costs to provide the equipment for defence, the electronics proportion of it within that total has more than doubled since 1949, and that will continue. (ibid., p. 92)

However, the electronics component varied greatly from one service to another: it was greatest in aircraft, big also in naval carriers, but relatively insignificant in an army tank (ibid., p. 98).

It is clear from contemporary evidence that already by the mid-1950s – no doubt partly as a result of World War II developments, partly as a result of the 1950s rearmament – some electronics firms were already developing the kind of special relationship with the GREs that had long characterized the aircraft industry, where during the interwar years 16 'family' aircraft firms and 4 engine firms had received preferential treatment (Postan 1952, pp. 435–6). On guided weapons, the ministry representative told a Commons committee in 1953, 'the basic research, almost without exception, is done in our own Establishments . . . The whole of the project work is done in industry. It is done under the scrutiny and surveillance of our Research Establishment' (GB SC Estimates 1953, p. 99). In industry the work was carried out by selected firms such as Hawkers (part of the

Armstrong-Siddeley group) and Vickers: a dozen major firms, out of a total of a thousand aircraft firms (ibid.) Before another committee in 1956, an Admiralty spokesman described similar procedures. Some 50% of total Admiralty R&D went on contracts to industry, and the procedure was to go to 'a contractor . . . who by reason of accumulated knowledge and experience is best able to undertake the work' (GB SC Estimates 1956, p. 11).

In the case of special equipment, the procedure was then to give first offer of production to the design contractor (ibid., pp. 83–4). However, the representative insisted, the Admiralty Signals and Research Establishment had an open policy on contracting; there was:

> a very large register kept. Anyone who has facilities, a decent bank account and some standing in the profession can ask to be included. Thereafter, it is entirely up to the Director of Contracts to select, say, half a dozen people from that register to tender to ASRE procurements specifications. (ibid., p. 89)

Similarly, the Underwater Weapons Establishment at Portland explained:

> 'Tender' is not quite the word we use for development contracts. What we have to find is a firm who is in a position to undertake the magnitude of the work likely to be involved without having to subcontract too much of it, and we also like to get some idea of their subsequent follow-up potential. I think it is true to say that we do not have too much difficulty. We have fairly good liaison with many firms. We like to know the potentiality of many firms, and then we boil them down to a reasonably small number. (ibid., p. 178)

Similarly, the Admiralty's Director of Naval Contracts explained that the selection of a firm

> would be put forward on the proposal of the technical department which was sponsoring the project, and we should have their advice as to the best firm to entrust that contract to. It would depend, for instance, on the capacity available in that field, the particular knowledge which one or other of the firms had in that technique, and so on; because, other things being equal, one would select a contractor who was an expert in that particular field which one wished to develop, and who would be more likely to bring it to a successful conclusion. (ibid., p. 202)

The firm would not necessarily contract to a fixed price; and a firm doing a development contract would be 'reasonably entitled' to a share of a subsequent production contract (ibid., p. 206).

By this time, clearly, certain electronics firms had a special relationship with the procurement agencies. The leading firms – English Electric, Edison Swan, General Electric, Mullard – were in constant touch with the Services Electronic Research Laboratory, through reading their reports and visiting them (ibid., p. 118); at the Admiralty Signals and Radar Establishment, valve manufacturers sat on official committees and in turn acted at hosts to official liaison meetings (ibid., p. 85).

Some firms, indeed, seemed to owe virtually their existence to government work. Solartron was founded in 1947 by two engineers in a disused stable, with a few hundred pounds of capital, to manufacture electronic testing instruments. Most of its early work was done on production and repair work under government contract. The firm moved into a small factory at Kingston upon Thames, manufacturing its first instrument in 1949, and then in 1951 to a larger factory at Thames Ditton. During the 1950s, it developed a wide range of electronic apparatus in order to escape from dependence on government work (Miller 1963, pp. 44–5). By the early 1960s, after financial restructuring, it had become a group of 14 companies which had established its main production base at Farnborough (ibid., p. 48).

So the essential features of the system were already in place by the mid-1950s. They consisted in a very close and intimate relationship between the Ministry of Supply (the precursor of today's Ministry of Defence), the GREs, and a few large suppliers; and in turn between the latter and a larger number of smaller specialized subcontractors. The nature of these relationships already seems to have locked the whole complex into a location close to the GREs themselves – that is, in the Western Crescent. To see how significant this was, we now turn to an account of the process at the present day.

Defence and industry today

The equipment budget for the Ministry of Defence for 1984–5 was £7 800 million, or 46% of the total defence budget. In addition the procurement executive spent over £700 million on associated R&D and management. The size of these outlays makes the Ministry of Defence (MOD) the largest customer of British industry, with over 10 000 contractors doing business with it at any one time. The MOD equipment programme accounted for 225 000 direct jobs and 276 000 indirect jobs in 1979 (Pite 1980); the 1984 Defence Statement, the MOD's annual review, estimated that despite large increases in real spending this had fallen to 125 000 direct jobs and 188 000 indirect jobs in industry, a result of the increased sophistication of the equipment (Dunne & Smith 1984; IPCS 1984).

More than 50% of national R&D is in fact channelled through the Ministry of Defence (GB Cabinet Office 1984). Of the defence budget 64% goes to private firms, 28% is spent within MOD and a mere 0.5% goes to universities. The intra-mural spending goes mainly on fundamental research within the GREs; the industrial spending on development. The MOD themselves say that in a number of industries – electronics, aviation control systems, marine technology – R&D would not function without them. Much of this goes to a small circle of large firms (GB House of Lords 1983, Gummett 1984).

Such industries are extremely defence-based: in recent years 45% of the aerospace industry's output, and 20% of the electronics industry's, has gone to the Ministry of Defence. In turn the defence budget is heavily biased towards these industries: aerospace receives over 29%, and electronics 21% of military spending, with a shift in recent years towards electronics (GB Ministry of Defence 1985).

Individual firms are even more heavily defence-dependent. Street and Beasley (1985) provide a detailed analysis of the defence work of 13 of the major companies involved in military contracts in the UK. Table 9.1, adapted from their report, shows the absolute value of defence work, both in total and for the Ministry of Defence, in each company relative to group turnover. Clearly, to all intents and purposes some of these major companies, such as British Aerospace, International Signal and Control, United Scientific Holdings and Westland, are just defence equipment developers and manufacturers. Although the MOD has as many as 10 000 contractors working for it at any one time, a large proportion of the total equipment expenditure goes to a small number

Table 9.1   Defence sales relative to group sales for selected major companies.

	Group sales £m	Defence sales £m	MOD sales £m
British Aerospace	2528	1980 (78)	902 (36)
Cambridge Electronic Industries	130	38 (29)	28 (21)
Ferranti	452	230 (51)	160 (35)
General Electric Co.	5600	1100 (20)	660 (12)
Hunting Associated Industries	179	95 (53)	95 (53)
International Signal & Control[a]	207	188 (91)	63 (30)[b]
Plessey	1218	448 (37)	190 (16)
Racal	816	320 (39)	95 (12)
Standard Telephones & Cables	1026	118 (12)	105 (10)[c]
Thorn EMI	2830	260 (9)	156 (6)
United Scientific Holdings	120	115 (96)	30 (25)
Vinten	23	12 (52)	1.5 (7)
Westland	296	250 (84)	180 (61)

Source: Adapted from Street & Beasley 1985.
Notes: Figures in parentheses show defence sales and defence sales to MOD as a percentage of group sales.
[a] Based in USA, but quoted in London.
[b] To US Department of Defense.
[c] Via prime contractors.

of firms. For example, if we sum the value of MOD contracts to the 12 UK companies in Table 9.1, they amount to £2 602 million or approximately 37% of the total MOD expenditure on equipment in 1983–4.

Although small firms are receiving relatively small amounts of money from the MOD, it can nevertheless be extremely important to them. A recent study of small high technology firms – admittedly in a period of rapid rises in defence spending – concluded that those with military contracts were growing 29% faster than those operating solely in civil markets (Forrester 1985).

This spending has a strong regional bias. The latest comprehensive figures for the regional distribution of defence procurement contracts (Short 1981a, 1981b) show that the South-East received 41.6% of the national total of contracts: five times as much defence expenditure in 1977–8 as any other region except the South-West (Table 9.2). Between them, these two favoured regions gained 58% of total UK contracts. The next most favoured region, the North-West, had a mere 9%.

On a per capita basis (Table 9.3), the southern regions again came

Table 9.2 Defence and regional expenditure, by regions, 1977–8 (£ million).

	Defence total	Regional (except defence)	Defence procurement contracts	Regional Aid Grants
South-East	2518.2	14022.2	1175.6	–
South-West	1107.2	3010.1	410.2	7.8
East Anglia	243.0	1274.0	111.7	–
East Midlands	382.7	2634.0	213.2	1.4
West Midlands	346.8	3720.8	189.2	–
North-West	382.1	5372.3	242.8	57.6
Yorks/Humberside	244.9	3759.2	61.9	26.7
Northern	242.1	2824.6	174.2	142.3
Scotland	497.1	5246.9	195.0	113.1
Wales	138.3	2481.2	25.6	74.4
N. Ireland	160.6	1797.9	27.9	50.1

Sources: Short 1981a and 1981b; Regional Statistics 1978, 1979.
Notes:
(a) Defence procurement contracts include: purchase of defence hardware for air, land and sea systems; clothing and textiles, liquid fuels, lubricants, etc.; research; and contract repair of ships and vessels.
(b) Expenditure for regions includes all government expenditure (central, local and nationalized industries) for the benefit of the region, but excluding expenditure for the benefit of the country as a whole, such as defence, prisons and foreign affairs.
(c) Regional Aid Grants include Regional Development Grants (paid under the Industry Act, 1972) and selective assistance grants. Note that the East and West Midlands together received £0.5 million in selective assistance grants which are not included in the table. For Northern Ireland the grant figures are for Investment and capital grants and Industrial Development Grants.

Table 9.3 Defence and regional expenditure, per capita, by region, 1977–8 (£).

	Defence total	Regional (except defence)	Defence procurement contracts	Regional Aid Grants	Totals	
	i	ii	iii	iv	i+ii	iii+iv
South-East	149.6	833.0	69.8	–	982.6	69.8
South-West	258.8	703.5	95.9	1.8	962.3	97.7
East Anglia	133.0	697.2	61.1	–	830.2	61.1
East Midlands	102.1	703.0	56.9	0.4	805.1	57.3
West Midlands	67.3	721.9	36.7	–	789.2	36.7
North-West	58.6	824.2	37.2	8.8	882.8	46.0
Yorks/Humberside	50.2	771.0	12.7	5.5	821.2	18.2
Northern	77.7	906.5	55.9	45.7	984.2	101.6
Scotland	95.7	1009.9	37.5	21.8	1105.6	59.3
Wales	50.0	896.3	9.2	26.8	946.3	36.0
N.Ireland	104.6	1169.5	18.2	32.6	1274.1	50.8

Sources: As for Table 9.2.

out at the top; the South-West received nearly £259 per capita, the South-East £150 and East Anglia £133. No other region received above the UK average of £112. At the other end of the spectrum, Wales and Yorkshire/Humberside each received only £50 per capita – one-third of the expenditure in the more prosperous South-East and under a fifth of that in the South-West (ironically the region bordering Wales). While the figures varied somewhat from year to year, the picture presented here was consistent throughout the mid-1970s (Short 1981a). In other words, one of the major parts of government expenditure tended to favour the prosperous regions both absolutely and relatively.

These figures quite swamp government regional aid, given primarily to firms in an attempt to equalize regional imbalances. Regional aid grants of course went to relatively depressed regions: in 1977–8, £142 million to the northern region, £74 million to Wales and £113 million to Scotland. But even on a per capita basis, if defence procurement contracts and regional development grants are combined, Table 9.2 shows that the resulting total expenditures strongly favoured the prosperous southern regions (see Lovering 1983 for a comparison of South Wales and the South-West).

In terms of employment, Dunne and Smith (1984) estimate that in 1977–8 423 000 jobs in the South-East were tied to defence, 237 200 of which were in industry (excluding Royal Ordnance Factories). The next highest defence-based employment was in the South-West (183 900, of which 82 700 were in industry). All other regions had estimated defence-related employment of under 80 000, and defence-industry employment of less than 50 000.

No comprehensive figures are available by which Short's (1981a and

1981b) analyses can be updated. However, the MOD has supplied this study with estimated regional percentage shares of procurement expenditure for 1983–4. These are presented in Table 9.4, alongside Short's (1981a) figures for 1977–8. The absolute figures for 1983–4 have been produced by multiplying percentages by the total procurement expenditure for that year. The comparison shows that the South-East has increased its share from 41.6% at 1977–8 to 54% at 1983–4, a dramatic increase over a short time period. This change is, however, consistent with what we know of general trends in defence expenditure, and particularly the increasing share of procurement expenditure going to the electronics industry, which, as we have seen, is very well represented in the South-East.

In real terms the defence budget rose by 23% or £8.2 billion from 1978–9 to 1983–4, compared to an overall growth of public expenditure planning of 7% and, for instance, a fall in the Department of Trade and Industry spending (which includes regional aid) of 29% (GB Treasury 1984). This relative shift to defence must favour defence-based industries and firms in the South-East and South-West. Within defence, expenditure on electronics rose by 119% and aerospace by 84.7% in the four years from 1979–80 to 1983–4, compared to an increase in total defence expenditure allocated to industries of only

Table 9.4 Defence procurement expenditure by region: estimates for 1977–8 and 1983–4.

Region	Expenditure 1977–8[a]		Expenditure 1983–4[b]	
	Absolute	Per cent	Absolute	Per cent
South-East	1175.6	41.6	3747.1	54.0
South-West	410.2	14.5	763.3	11.0
East Anglia	111.7	3.9	208.2	3.0
East Midlands	213.2	7.5	346.9	5.0
West Midlands	189.2	6.7	277.6	4.0
North-West	242.8	8.6	763.3	11.0
Yorks/Humberside	61.9	2.2	138.8	2.0
Northern	174.2	6.2	138.8	2.0
Scotland	195.0	6.9	416.3	6.0
Wales	25.6	0.9	69.4	1.0
N. Ireland	27.9	0.9	69.4	1.0
Total	2827.3	100.0	6939.0	100.0

Sources:
[a] Short 1981a.
[b] Ministry of Defence. In providing these figures, the Ministry point out that these estimates are based on 85% of contracts and that the missing contracts might affect the regional distribution, possibly depressing the South-East share below that shown here. Absolute figures have been calculated from the percentage figures supplied.

78% (GB Ministry of Defence 1985; see Table 9.5). All this makes it unsurprising that high technology industries have continued to grow in the South-East and South-West in the early 1980s. In short, these regions have been favoured by a large and rapidly expanding segment of government expenditure, while regional aid has fallen.

However, there are also sharp differences in the type of defence equipment procured in each region (Law 1983). The Ministry of Defence contract office for clothing is in Leeds – but spin-offs to other civilian uses, or foreign exports, are less likely than for, say, sophisticated electronics. Similarly, spin-offs from plants producing long runs of a piece of equipment may be less than from plants concerned with research, development and prototype manufacture. Spin-off of military technology into civilian uses is more likely in small firms (GB NEDC 1983, Forrester 1985), where the entrepreneur or employees may be working – or in close contact – with both the 'civilian' and 'military' products or sides of the business, than in a branch of a large firm devoted entirely to military products. Anecdotal evidence suggests that the South-East has more higher qualified workers engaged in product development, who may be more likely to start their own small firms; and in these small firms there may be greater spin-offs to civilian uses.

A number of large firms have their production facilities in less prosperous regions, but retain research, development and prototype production in the home counties. However, such functional splits are not common to all firms. For instance, Racal's company policy means that no 'division' is larger than 600 employees. Hence the full range of functions is carried out by a single 'division', usually in a single town. The majority of such divisions are just west of London, with another near Tewkesbury, close to Malvern. Such a company structure would result in fewer production jobs in development regions – though, if such a division did locate in a development area, it might beneficially bring a full range of functions.

But large firms in large plants, with highly skilled employees, are likely to generate agglomeration economies in the form of skilled labour pools and suppliers, which might attract or benefit firms with no defence links. Around even a few large defence-orientated plants, small subcontracting firms and new firms set up by ex-employees may grow up, leading to a cumulative advantage for the area (Breheny & McQuaid 1984). Given the concentration of electronics firms in the South-East and South-West, it is likely that these spin-off and agglomeration effects are largest there.

We have demonstrated fairly conclusively that there is a distinct regional and, indeed, subregional bias in the awarding of defence contracts. Given the importance of this expenditure in the development of the highly prized high technology industries, this bias needs to be explained. We have part of the explanation already. We have traced the historical tie-up between the electrical, and then the electronics industries, and the DREs; clearly, this historical link goes some way

Table 9.5 Defence expenditure in the United Kingdom: estimated allocation by commodity group (£ million).

SIC (1980)	Activity Headings	1979–80 Total	1979–80 Share (%)	1983–4 Total	1983–4 Share (%)	1979–80 1983–4 change (%)
solid fuels	111–120	8	0.02	10	0.01	25.0
petroleum products	140	427	10.71	741	10.44	73.5
gas, electricity & water supply	161–170	111	2.80	182	2.56	63.9
ordnance & small arms & explosives	256 (part), 329	311	7.80	575	8.10	84.9
other mechanical & marine engineering	320–328	221	5.54	410	5.77	85.5
data processing equipment / other electrical engineering	330 / 341–348 NES	128	3.21	211	2.97	64.8
electronics	344,345	699	17.53	1531	21.57	119.0
motor vehicles & parts	351–353	133	3.33	180	2.53	35.3
shipbuilding & repairing	361	331	8.30	424	5.97	28.1
aerospace	364	1120	28.10	2069	29.15	84.7
instrument engineering	371–374	101	2.53	123	1.59	21.8
food	411–429	93	2.33	113	1.59	21.5
textiles, leather goods & clothing	431–456	78	1.95	93	1.81	19.2
other production industries	111–495, NES	119	2.98	290	4.08	143.7
other industries & services	NES	104	2.61	144	2.02	38.4
Total		3986	100.00	7096	100.00	78.0

Source: Adapted from Table 2.5, GB Ministry of Defence 1985 Vol. 2.
Note: NES – not elsewhere stated.

towards explaining the current geographical pattern. However, we have not so far explained why this geographical pattern has been sustained, and if anything strengthened, during the postwar period. Reading the American literature, one would not expect the close geographical link to have been kept up. Commenting on the spatial distribution of the defence industry in the United States, Markusen (1984) states that: 'physical orientation [of the contractors] to the customer's headquarters [the DOD] is not important' (p. 10). Why, then, in the UK have defence industry companies continued to locate so close to their customer's headquarters (Whitehall and South-East DREs)? One answer might be in the nature of the procurement process itself in the UK.

The defence equipment procurement process

The procurement process has two central features: lack of competition, and intrusion of political factors (Angus 1979, Greenwood 1982, Pearson 1983). On the latter, there is abundant anecdotal evidence. The Secretary of State for Defence admitted in Parliament that when ordering two new warships, the cheapest solution was not adopted, but 1700 jobs were maintained at one yard (*The Financial Times*, 29 January 1985). The Chief Secretary of the Treasury successfully lobbied for establishment of the Military School of Music in his constitutency, though the MOD admitted that other locations would have been cheaper (*The Guardian*, 30 June 1985).

Quite probably, such interference might help economically distressed areas. But lack of competition almost certainly has the opposite effect. Only 20% (by value) of contracts, excluding subcontracts, were awarded through competitive tendering in 1982–3. This implies that costs are not the most important factor in awarding contracts; in fact, a recent study found average savings of 30% following the introduction of competition (GB Ministry of Defence 1984b). The critical consideration is the technical capability of the contractor, but this may be difficult to determine unless a firm has a long-standing relationship with the MOD. And, though the 1984 Defence Statement promises to increase competition, it also stresses that there are inevitably some limits to which competition can be introduced: the country needs a strong indigenous defence/industrial base, and there is a limited number of British suppliers of certain advanced equipment (GB Ministry of Defence 1984b, p. 17).

Given the lack of competition and limited possibility of introducing it, what factors are important in awarding contracts? And do they have a spatial bias? The Procurement Executive of the Ministry of Defence is responsible for purchasing equipment and other goods and services for the Navy, Army and Air Force. Non-specialist building work, furniture, stationery and office supplies, plus administrative and scientific (not operational) computers, however, are purchased through other government departments (GB MOD, n.d.). The three main ways in which a firm can sell to the MOD are: as a prime contractor; as a subcontractor to a single prime contractor; or direct to local major defence establishments, although usually only government defence research establishments (DREs) have powers for contracts concerning more than day-to-day requirements. Figure 9.1 shows the location of major DREs and what the MOD calls 'contact points' – administrative centres where prospective contractors have to negotiate with the Ministry. These 11 DREs (listed in the last chapter), subsequently rationalized to 7, have a particular status; that of 'local purchasing powers'. This means that staff at the establishments have considerable autonomy in spending MOD funds. The obvious feature of Figure 9.1 is that all but one of the contact points (Leeds, where contracts for uniforms are dealt with!) and one DRE (Malvern, on the very edge of the South-West region) are in the South-East or the South-West.

The majority of prime contractors are also located in the South. Of the 57 UK-based contractors who were paid £5 million or more for defence equipment in 1981–2, 43 had their head offices in the South-East, and 4 in the South-West (GB MOD, n.d.). Admittedly, parts of the contracts and subcontracting might be carried out elsewhere. But these firms generally had most branches, particularly R&D and higher level functions, in the South also. The large, technologically complex projects tended to go to a few contractors: members of the 'inner club' (Angus 1979). Presumably, local subcontractors in the South find it easier to get and maintain contacts with these major firms than would equivalent small firms in another part of the country, especially for non-standardized or short production-run products, because they can readily resolve difficulties and make minor modifications to products. However, perhaps the strongest reason for the spatial bias in defence contracts is the equipment development process.

There is consistent pressure to produce ever more sophisticated weapons systems and communications and other military equipment. Recently most systems and much equipment have incorporated advanced electronics (see for instance Blackaby 1983). These systems are often developed by high technology firms in close association with researchers in government defence research establishments and with officials of the MOD procurement executive. Engineers and scientists from the firms work closely with their counterparts in the DREs to develop new equipment, install and test prototypes and even to guide the DRE scientists along certain lines of new developments. While the final decision on the development and adoption of any new ideas may ultimately rest in MOD headquarters, the advice of experts in the DREs and their committees doubtless carries much weight; it also lays out the options which may limit the choice of Whitehall.

On the nature of this process, there is abundant evidence from Parliamentary inquiries from the 1960s onwards. A Select Committee in

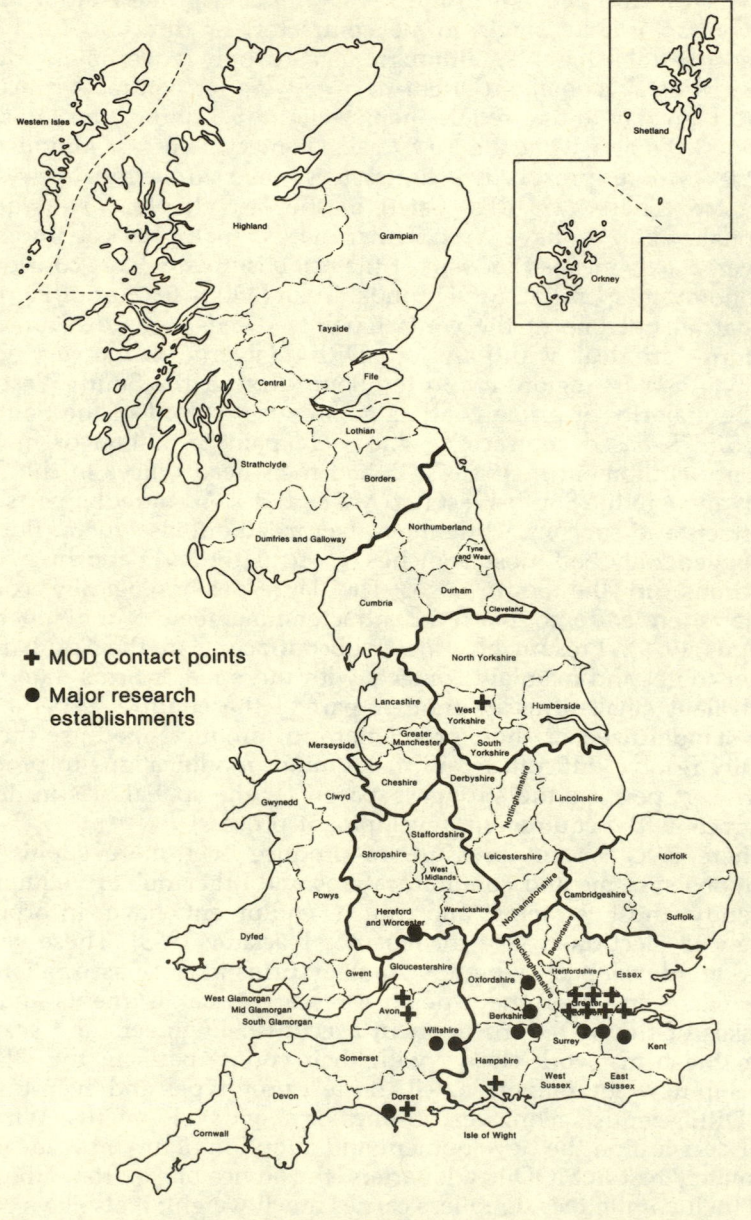

Figure 9.1 Ministry of Defence GREs and contact points, 1983.
Source: Ministry of Defence (n.d.).

1968 was told by the Ministry of Defence that in industry generally about 37% of all R&D was government financed, but that in aircraft and electronics the figure was 72% (GB SC Science 1969, p. 72). The president of the Electronic Engineering Association informed the Committee that 'We think that such technological position as we have today in the world is based primarily on military spending to which the establishments and industry and a lot of other agencies have contributed' (ibid., p. 187). It might be an expensive way of keeping in the technological vanguard, he agreed, but there was no alternative (ibid., p. xv). The Committee concluded: 'Industry, particularly in the fields of electronics and aerospace, is extremely dependent on research contracts' (ibid., p. xxv). Thus it was, in the words of the industry representative, that:

> Our industry deals with Christchurch, S.R.D.E., it deals with R.R.E., it deals with R.A.E., it deals with the Admiralty, A. and A.E.E., R.A.R.D.E., all of those which support the military machine we are continually in contact with. (ibid., p. 194)

Plessey, who told the committee that 50% of their R&D was done for government, frankly told the Committee that 'removal of Government R&D for the electronics industry would have a profound effect on the competitiveness of the industry' (ibid., p. 15). They were equally forthcoming about the relationship which existed between the establishments and the companies; it was an 'old boy network . . . based on long association and mutual trust' (ibid., pp. 64–5). This relationship existed between the staffs on the two sides, and it was based on professional association in institutions like the Institution of Electrical Engineers and on a daily professional level in the business they did (ibid., p. 17). Plessey's R&D organization had grown from 200 R&D personnel to over 7000, based at West Leigh, Havant and at nearby Roke Manor; the percentage of R&D personnel engaged on defence work was suppressed from the public record (ibid., p. 20).

Marconi, another major government contractor, told the same story. After the war, and particularly during the defence buildup of the 1950s,

> a number of industrial engineers emerged who had built up close working relations with the Establishments and who, by reason of wartime involvement in defence problems, and close contact with the Services in the post war era, had come to supplement in a most important way the official defence scientists of the Establishment and the defence services. (ibid., p. 379)

These relationships, he reported, were especially good with the Radar Research Establishment; they were also good with the Royal Aircraft Establishment and the Signals Research and Development Establishment (ibid., p. 385). GEC and AEI had similar laboratory teams, he reported (ibid., p. 391).

Similarly, from the government side, the Controller of Research at the Ministry of Technology confirmed that at the Radar Research Establishment:

> We have an open door to those firms we are dealing with in the industry. They have normal access. For instance, all the major firms in the electronics industry are in and out of RRE on the research programme. This is on a day-to-day basis. (ibid., p. 162)

These links, however, were essentially with the big industrial research laboratories, for only there 'you do find people who talk the language, understand what you are after, and what is involved in the physics field in getting there. So it is mainly with big firms that our contracts are' (ibid., p. 163).

These cosy relationships, once established, were not easy to change. Slowly, and partially, they did; by 1981, a Plessey representative was lamenting to a Commons committee that the growth of competitive tendering had brought in new people, lacking the right experience (GB Defence Committee 1981, p. Q752). But during the 1950s and 1960s, given these relationships, it was only to be expected that the major companies would cluster around the GREs. True, not all the leading electronics companies were concentrated in one area; Plessey were at Havant, while Marconi's research facilities were at Great Baddow outside Chelmsford (GB SC Science 1969, p. 381). Nevertheless, it seems clear that for newer entrants to the field a location near the establishments might make a crucial difference in establishing the necessary contacts. And, despite the anomaly of RRE, there is no doubt that the centre of gravity of the establishments lay in the Western Crescent – especially in its southern part.

The interview evidence

Our interview survey, already reported in Chapter 5, provides abundant evidence of the importance of such personal contacts between staff of DREs and firms. At the eastern (Berkshire) end of the Corridor 44 firms were asked about their links with DREs and the importance of defence contracts. (They did not include GE/Marconi, perhaps the biggest defence contracting firm in Britain.) Table 9.6 shows the estimated proportion of contracts (including subcontracting) related to the MOD. Exactly half of the plants (each representing a different firm) claimed significant defence contracts (either as prime contractor or subcontractor), and eight of the ten largest plants had such contracts. Interestingly, six of these eight suggested that the location of DREs had some importance, while only five of the nine medium-sized firms thought so. This may reflect the closer ties of large firms to DREs and the greater technical sophistication of their products. Respondents often cited a pool of skilled labour as a feature

Table 9.6 Defence sales as a percentage of output for high technology plants.

Size of plant (workers)	Per cent value of output for defence contracts and subcontracts				
	Not significant	5–20	20–50*	50+	Total
100+	2	3 (2)	1 (1)	4 (3)	10
20–99	10	1 (1)	4 (2)	4 (2)	19
1–19	10	1 (0)	3 (1)	1 (0)	15
Total	22	5 (3)	8 (4)	9 (4)	44

Note: Figures in parentheses indicate plants mentioning that the location of DREs was of some significance to them.
* Three firms answering 'high' or 'considerable' included here.

of the location; and this pool is provided in part by the presence of large British defence firms. In central Scotland, similarly, Ferranti's defence plants have provided a pool of skilled labour for foreign firms moving into the region (Firn & Roberts 1984).

Many firms reported that only a small proportion of their output went to MOD contracts. But this can be misleading. In one case, the MOD work paid for most of the basic research carried out by the firm; and this was essential to improve the civilian products which made up the majority of sales. While there has been much criticism of the lack of such military–civil spin-offs in Britain (Udis 1978, GB NED 1983, Gummett 1984), in this firm they were facilitated by three factors. First, the managing director was an engineer with intimate technological understanding of how ideas from military research could feed into civilian products. Secondly, his firm was small so that all personnel had continual contact with one another. Thirdly, the civilian and military technologies were similar enough to allow easy transfer of developments, and as both were extremely advanced in their fields they could continue to be developed together. The key locational criterion for this firm was to be within two hours' drive of the several DREs with which it dealt. This had inhibited a move to a development area when the company first started.

Further evidence comes from another small company which gained a contract with a DRE to combine two types of equipment, which it was already manufacturing, into a new integrated unit. The contract provided enough funds for the firm to survive for a year. Since then the product has been adapted and sold to the civilian market and the firm has rapidly expanded. Again, except for this one year, only a small share of output was military; yet the basic research funds were critical to the firm. These two cases also highlight a more general observation: defence funds may not only affect the location and subsequent expansion of a firm, but also the way in which it develops.

An interesting perspective on the attitudes of the larger prime-contractor firms comes from three managers in one of them. They felt that if their plant moved a further 100 miles (161 km) away from London, visits from the military would fall by half. This may be exaggeration; the point is that they believed it, and that this influenced their location. Asked what would happen if the relevant DREs and the MOD procurement section moved to Liverpool, they replied that research, development and prototype development would have to move closer to Liverpool; much of the rest of the firm would probably follow. They added that they would go not to the city but to somewhere attractive close by, like Cheshire. This illustrates that environmental factors may be very important, but only once the general location is chosen.

Further anecdotal evidence from the survey confirms the view in the literature concerning the great importance of face-to-face personal contacts (Törnqvist 1968, Goddard 1973, Goddard *et al.* 1983). But here, these favour not the city centre, but the areas immediately around the DREs – which incidentally allow easy access to Whitehall when necessary. These contacts are greatly strengthened by the large-scale transfer of ex-MOD personnel to posts with defence-based firms, and occasionally the reverse. According to the Treasury and Civil Service Committee in 1984, some 1 404 MOD officers applied for business appointments with companies, and only 14 of these were refused. In one case the chairman of a defence-based firm in Cambridge (United Scientific Holdings) was appointed head of government defence procurement, while his old job as chairman of the company went to a former Permanent Secretary at the MOD (*The Financial Times*, 20 December 1984. Top-level MOD officials and DRE staff are certain to live in the South-East and will wish to remain there when joining defence firms. In so far as they are then effective in gaining continued or additional work for their new employers, firms in this region will have a considerable further advantage.

The reverse is also true, of course. An engineer from a north of England high technology firm, with considerable defence business, was asked if his location was a disadvantage. He replied that it certainly was, since it was not possible to 'call in for coffee or lunch' and have informal discussions with personnel at DREs.

The basic model of defence contracting, particularly in high technology, thus has a first stage in which face-to-face contact between DRE and Whitehall personnel and those of firms is vital. Here the firm brings suggestions and responds to government ideas concerning the development of new equipment. It must be remembered that as often as not the contractors will be telling the military what the military wants rather than waiting for orders. Lobbying is thus an important part of the contracting process. Through more informal meetings, as well as through organized committees, further ideas can develop and prototypes can be designed and/or produced at either the DRE or the firm. After the prototype has been installed at the DRE or a related testing and trial location – requiring more personal exchanges between the different parties – the contract for full production (if any) would be open, but the initial prototype developer would be most likely to get it. Long-run production may then take place elsewhere in the country. Clearly this is a simplistic model, but it does highlight the importance of face-to-face contacts between personnel of the firm, the DREs and Whitehall. If the R&D sections of prime-contractor firms are located near particular DREs, many subcontracts may also go to small local firms and to other large defence firms locally.

Conclusion

Defence expenditure is thus vital to the spatial structure of the economy because of its size – over 5% of GNP – and because it represents the largest single market for British industry. Total defence expenditure favours the South-East and South-West in both absolute and per capita terms. While expenditure on bases and installations might be dictated primarily by strategic considerations, this is not the case for the bulk of defence equipment procurement contracts. These are placed mostly with private firms, but the majority go to the South – far outweighing regional aid. Recent increases in defence spending, channelled more and more into electronics, have proved resilient to recession and to expenditure cuts, and they benefit particularly the South-East and South-West.

This spatial bias, we have argued, was not simply a case of historical accident. Rather, defence-related firms – especially the R&D functions – located and expanded in the South-East because of the importance of face-to-face contacts with staff at government defence research establishments and in Whitehall. These DREs and the procurement contact points of the MOD were situated along a crescent around west London and along the broad Bristol to London axis. The recent reorganization and concentration of DREs within this area is merely likely to consolidate the advantages of the most prosperous part of the country.

This dominance is unlikely to alter radically. Defence budgets are rising both in Britain and in the United States – and the Strategic Defense Initiative should lead to a big increase in high technology R&D spending with industrial spin-offs, some perhaps coming to Britain. (A group of British firms that visited Washington late in 1985 to try to obtain non-weapons defence contracts, however, came away empty-handed.) American evidence suggests that defence expenditure tends to increase at above the rate of inflation – although, there, the 1985 Gramm-Rudman Act could lead to future squeezes. In 1985 it was suggested that the Ministry of Defence might try to shift procurement contracts to Assisted Areas; but the fact is that the existing expertise is still in the South, and while it remains there, the technological imperative will continue to place the crucial high tech work there. So a central dilemma will remain for regional policy.

10

The role of planning

Public policy, we have seen, had a profound effect on the growth of the Corridor through the establishment of the GREs and their relationship with the defence contractors. It had yet another, through the impact of land-use planning policies. But this impact was paradoxical: through negative controls – restrictions on new industry, restraints on new urban development – it maintained the area's traditional agricultural character, but thereby made it ever more attractive to high tech enterprise.

The wartime foundations of policy

We find the origins of this paradox in World War II. Then, critical planning reports – the Barlow Report of 1940 on the distribution of the industrial population, the Abercrombie Reports of 1943–4 on the planning of London and its region – laid the foundations for policies that, with relatively little deviation, would be followed for the next 40 years (GB RC Distribution 1940, Forshaw & Abercrombie 1943, Abercrombie 1945). The critical point is that they were developed in reaction to the trends of the last years of peace, but against a background of total war.

The 1930s, as seen in Chapter 6, were a time of massive expansion of the high tech industries of that time, and of their concentration in north-west London. Many of these firms had begun to cluster along the major arterial roads to the west and north-west – Great West Road, Western Avenue, Barnet By-Pass – and into new industrial estates in places like Park Royal and Hendon, or, farther out, Slough or Elstree. Above all, this tendency was noticeable in west London:

> The extent to which industry here has followed the main railways and roads is very noticeable . . . the Great West Road industrial concentration, extending westwards from the junction of the Great West Road with the Chiswick Road . . . is the most outstanding example of an 'industrial ribbon' in the London Region. (Abercrombie 1945, p. 43)

There were good reasons, as Abercrombie himself recognized, for this clustering: the new arterial roads and rail facilities, connecting the firms both to the west and to the West End, commercial heart of London and therefore main market; the available land, ready-serviced in industrial estates; the area's 'clear atmosphere' (ibid., p. 42).

Yet, as Abercrombie and Forshaw had complained in their *County of London Plan* (1943), such decisions were taken neither by the planner nor by the industrialist, but effectively by the industrial estate speculator; and 'however efficiently the estates may be planned, they were not always chosen with due regard to the needs of the region as a whole' (Forshaw & Abercrombie 1943, p. 95): in their wake – so it was argued – came unsuitable residential developments, traffic congestion, the merging of ribbon developments into continous sprawl, loss of valuable agricultural land and deterioration of the environment (Abercrombie 1945, p. 13). The Great Wen, already the largest city in the world (with 8.5 million people, its peak population, in 1939) would spread to envelop the precious agricultural land, and irreplaceable landscape, all around. 'This, at all costs', Abercrombie and Forshaw intoned, 'must be prevented' (Forshaw & Abercrombie 1943, p. 92).

Thus the argument of the reports was a reaction to the 1930s. But their force, and in particular their official and popular reception, reflected the exceptional circumstances of the 1940s. External threat had conditioned people to accept a level of direct government control unimaginable before, unthinkable even today. For military reasons, government took over the output of whole industries, directly invested in research and development, constructed shadow factories outside London, and directed existing factories to relocate to safer areas – especially west of London (Foley 1963, Hall *et al.* 1973). It was then easier to accept the radical conclusion of the Barlow report: that, partly for strategic reasons, firms should be prevented from moving into Greater London by a system of official industrial licensing. After all, it was not the radical Attlee administration, but the predecessor wartime Coalition government, that passed the Distribution of Industry Act of 1945, giving this power not just over London (as the Barlow majority had recommended) but over the whole of Britain.

Similarly, bomb and rocket damage made it easier to accept that major physical reconstruction would be needed at war's end, and that this could not properly be accomplished without large-scale planned overspill of people and jobs along lines long argued by the Garden Cities and Town Planning Association. New towns, green belts and protection of the countryside passed from the realm of the radical and utopian to that of the sensible and practicable. All were embodied in Abercrombie's bold vision for Greater London. London's outward growth was to be halted by the green belt; on the far side of this, new industry should be strictly controlled and concentrated in planned satellites, new towns and expansions of existing towns; these should be tied to each other, and to London, by new high-speed roads and electrified railways (Forshaw & Abercrombie 1943, pp. 33–5, 62; Abercrombie 1945, pp. 26, 58, 71–3, 84, 127). The result was a spatial pattern for the entire region which included the existing built-up area, the surrounding green belt, and an Outer Country Ring – a combination of open countryside, new and expanded towns – which was to provide the major reception area for the overspilled population (Abercrombie 1945, pp. 8–9, 33–5, map opposite p. 66). And this was to be achieved by a combination of two measures: controls on industrial location in the London region, incentives for transfer elsewhere, as contained in the 1945 Distribution of Industry Act; and physical controls on urban growth, to be made possible by the new powers given to local authorities by the 1947 Town and Country Planning Act.

One point was especially significant for subsequent history. Abercrombie's emphasis was not merely on containment but on self-containment: though existing cities and planned towns were to be linked by road and rail, 'unnecessary' travel – particularly long-distance commuting – would be discouraged, by putting people and jobs close together within each urban area, and by stressing separation rather than inter-connection between them. So Abercrombie's world was a static, immobile one: without growth (an error of prediction he partly took from the demographers, partly based on his own overconfidence in industrial controls) and without much movement either.

Soon, it emerged that these assumptions had badly underestimated the power of technological and economic change, and had as badly overestimated the planner's ability to control them. Irresistible forces met an immovable object; the ensuing collision forced a major policy shift. Particularly, a conflict emerged between the aim of promoting economic development, and that of conserving the environment: a conflict that found central government departments on one side, local planning authorities on the other. And nowhere was this more evident that in the Western Crescent.

Implementing the plan:
the Western Crescent

Immediately after World War II such issues were but dimly perceived. The main concern of both central and local government was not whether Barlow and Abercrombie were right, but how best their recommendations could be implemented. Even then, however, conflicts and trade-offs soon emerged.

At local level, during the 1950s the nine county planning authorities within the *Greater London Plan* area all adopted Abercrombie's suggestions in their own statutory development plans. But the vision of a single planner was replaced by 'a multiplicity of sundry government authorities, some decidedly powerful in their own right' (Foley 1963, p. 79); a coherent, cohesive policy for a large area was 'factored' according to the different perspectives and priorities of the local authorities (Hart 1976). Neither the principles nor the broad regional pattern was challenged; but details were changed. Some specific proposals were scaled down; a few rejected entirely; several relocated.

Specifically, the Development Plans excluded any expansion in the Green Belt – with the important exception of Slough – and limited the amount of expansion farther out, which was concentrated in a few selected sites for people 'decanted' from London (Hall *et al.* 1973, ch. 5). Great care was taken to preserve the character of historic towns like Windsor, Ascot and Henley, and the high-quality farmland of the Thames Valley. The Board of Trade proposal, that Slough should become a key overspill location for 100 000 people and 40 000 jobs from London, was received with caution. And Abercrombie's suggestion for a new town for 60 000, at White Waltham, west of Maidenhead, was rejected by the government – ironically, given its authorship – on the ground that its proximity to the existing Western region main line and the future M4 would encourage ribbon development from Slough to Reading (Abercrombie 1945, pp. 215, 409; Foley 1963, pp. 31–2; Hall *et al.* 1973, pp. 458–60).

Farther out, the local planners faced a major problem in the continued growth of Greater Reading. In the late 1940s planners in the three responsible authorities – Berkshire, Reading, Oxfordshire – established a Joint Advisory Committee to deal with its overspill. By the early 1950s, 'urged on by the Ministry which took the lead in formulating a positive suggestion', they agreed a Blue Line policy placing a limit on urban development, which they incorporated in their plans (Hall *et al.* 1973, p. 465; Reading Borough Council 1983). But a difference was already appearing: the Board of Trade thought that Reading's excellent communications would make it a suitable reception area for London industry. Central government, it seemed, would countenance growth in a few selected, compact sites; local authorities were more resistant to growth of any kind, whether of jobs or homes.

This growing divergence was highlighted in the search for new town sites. After rejection of White Waltham, Bracknell was designated in June 1949 – accepted locally, with some rumblings, as a necessary evil. Yet after this, even while the local planners were running their Blue Line round Reading, the town, together with Newbury and Basingstoke, was being considered 'more or less seriously' as a further new town candidate (Cullingworth 1979, pp. 112–13). A head-on central– local confrontation was avoided, for the proposals remained just that: no new towns were designated in England between 1950 and 1961. A drop in birth rate meant that the pressure was off – for a time.

The 1960s: the era of regional studies

But not for long. The year before the 1961 Census, an influential paper, from a Ministry of Housing and Local Government official, already concluded that population and employment in the South-East were growing far faster than the Abercrombie-based Development Plans had assumed; and that the forward predictions were therefore far too low. Geoffrey Powell's analysis showed that although Greater London's population had stagnated and was even beginning to decline, the ring on the far side of the Green Belt – Abercrombie's Outer Country Ring, now officially retitled the Outer Metropolitan Area (OMA) – was now the most rapidly growing part of the country. During the 1950s this ring, extending from the inner edge of the Green Belt to some 40–50 miles (60–80 km) from central London, had one-third of the net British population growth (Powell 1960). Thus, while the green belt had stopped London's physical growth, it had done nothing to stem the capital's functional growth into a city-region of enormous size.

The 1961 Census results confirmed Powell's analysis; and, in government circles, concern grew that the planning system of the region was unprepared for the new growth. Thus, just as wartime problems triggered the Abercrombie plans, so did the challenges of peace stimulate a new wave of plan-making that was to continue for more than a decade. Three major reports resulted: *The South East Study 1961–1981* (GB Ministry of Housing and Local Government 1964); *Strategy for the South East* (GB South East Economic Planning Council 1967); and the *Strategic Plan for the South East* (GB South East Joint Planning Team 1970).

These plans reflected a new awareness in Whitehall, not yet shared by the local authorities, of the implications of growth. Powell had suggested, and the 1964 *Study* underlined, that population growth in the South-East arose mainly not through in-migration from less prosperous regions, but from unexpected natural increase among the region's own young people. In Berkshire (including Reading, then a separate planning authority), the *Study* showed that the 1961 population of 522 000 would by 1981 grow by no less than 146 000 – 117 000 of which would be natural increase (GB Ministry of Housing and Local Government 1964, p. 86, Table 7). This had profound implications: for new schools, new health care facilities, and above all new housing.

Similarly, the economic trends were changing. Not only – contrary to plan – was employment continuing to grow rapidly in the South-East; the character of that employment was shifting. Barlow and Abercrombie had stressed control of manufacturing. But it now became evident that traditional smokestack manufacturing was stagnant or declining, while new kinds of manufacturing, and above all new kinds of service activity, were rapidly growing. Early high tech was particularly important in this respect. From 1959 to 1962, radio and other electronic apparatus added 29 900 jobs (20%), making it the fastest-growing factory sector in the South-East; within the service sector, air transport grew by 8 100 or 24%; postal services and telecommunications by 10 000 or 13.8% (GB Ministry of Housing and Local Government 1964, pp. 138–9, Table 21). And it was becoming evident that offices and related developments were clustering in a relatively small number of major concentrations, one of the most important of which was greater Reading.

The *South East Study*, produced in 1961–4 by a group of civil servants in the Ministry of Housing and Local Government led by Geoffrey Powell, recognized these facts of growth. It opened: 'There are expected to be at least 3½ million more people living in South East England by 1991; it might prove more' (GB Ministry of Housing and Local Government 1964, p. 1). This was notwithstanding a policy assumption that the new *Study* shared with the old Abercrombie Plans: that government policy would continue to channel away economic growth to other parts of the country and that most of this growth took the form of traditional manufacturing (ibid.). The new feature was the recognition of the region's dynamism, which planning could not control, only guide. Twenty years after Abercrombie, the *Study* marked the end of one era and the start of another (Hall *et al*. 1973, p. 470). True, in the resulting policy prescriptions there were elements of continuity both in the underlying principles and in their physical expression. But there was a new feature: the old image of a static city region, composed of small and largely self-contained units, was giving way to a vision of larger, much more interactive, city-regional entities.

As the *Study* noted, 'To be effective these centres would have to be large and strong' (GB Ministry of Housing and Local Government 1964, p. 53). To guard against further spread of London's commuter ring, they should be 60 miles (100 km) and more distant. There should be substantial new cities at Southampton–Portsmouth and Bletchley. And, harking back to the new town suggestion of a decade before, Newbury should grow from 20 000 to 98 000 by 1981, and eventually to a quarter of a million (ibid., p. 73, Table IV). But, as the authors observed:

There is no existing industrial potential in the area, but the natural advantages of its position should make up for this; again, for this reason any major development would have to take the form of a new town. One drawback is that much of the surrounding area contains farmland of high quality. (ibid., p. 74)

As well as new cities, there were 'big new expansions' and 'other expansions'. Under the first head, Swindon would grow from 90 000 in 1961 to 154 000 in 1981. Under the second, Reading would expand from 120 000 to 166 000, though perhaps over a longer span (ibid., p. 73, Table IV). And to balance these, the *Study* stressed the need to maintain the integrity of the green belt and to develop high-speed road and rail links (ibid., pp. 33, 91, 58–64).

The *Study* thus represented a conceptual revolution. Containment of London remained an objective; but there was a new stress on accommodating continued growth. And the older image of self-contained new and expanded towns was replaced by a notion of linkages within and between urban complexes, forming a vast functional city-region several thousand square miles in extent. The resulting pattern would be mutually beneficial both for London and for the surrounding counties. Not only could everyone have their cake and eat it too; everyone could have a larger slice.

But that is not how the counties saw it. At precisely this time, they were beginning their required quinquennial reviews of their own development plans and also their more detailed town map proposals. And these plans did not always agree with the recommendations coming from the centre; indeed, sometimes they were in direct conflict. In particular, while the *Study* urged the counties to identify land for urban growth, and only then to make limited extensions to the green belt (ibid., p. 93), the counties themselves were bent first on major extensions of the green belt – the so-called interim green belt – precisely in order to stop further development.

The problem was that the local plans, once agreed by the minister, had statutory force; the central studies and plans were merely advisory. True, the centre had power to build new towns, motorways and airports; yet these kinds of direct intervention were minor compared with the land-allocating capacities of the counties. And, although central government could always modify development plans and overturn development control decisions, this power could be used only sparingly. A balance had to be struck: between growth and conservation, between central and local government power. After the *Study* it was doubtful whether this balance had been achieved – and, consequently, whether the *Study* was capable of rapid or full implementation.

But these were years of rapid institutional change. Early in 1966, as part of the Wilson government's regional planning initiative, the South East Economic Planning Council came into existence, advising the

Secretary of State for Economic Affairs, George Brown. In November 1967 it published its *Strategy for the South East*, which developed the principles of the 1964 *Study* into a fully fledged physical structure based on discontinuous corridors of growth radiating outwards from London. The Western Crescent would contain two of the largest in the country: a major corridor straddling the M3 from London to Southampton–Portsmouth, a minor corridor along the M4 from London to Swindon (GB South East Economic Planning Council 1967). Joining these, 40 miles (70 km) west of London, was an area of strong growth bounded by Reading, Basingstoke, Aldershot and Bracknell, which, so the Council argued, was one of a number of places needing special further study through a partnership between central government and local authorities.

At a conference organized by the Town and Country Planning Association in January 1968, local reactions to the *Strategy* ranged from cool to hostile. This was crucial, since the local authorities were the chief – indeed, apart from the new towns, the sole – implementation agencies. Central government and its advisers were at loggerheads with grassroots local opinion, and nowhere more so than in the growth areas of the Western Crescent. To help resolve the impasse, in 1968 the government launched yet another planning study – this time, jointly commissioned by the Ministry of Housing and Local Government, the Economic Planning Council, and the Standing Conference of local planning authorities. It would be led by Wilfred Burns, newly appointed Chief Planner at the ministry; it would be produced by more than 40 central and local government planners and outside experts. It would, in the words of the delicately drafted terms of reference, start from the Planning Council's *Strategy* but would also take account of the work of the Standing Conferene (GB South East Joint Planning Team 1970, p. 1); it would provide a framework for the new Structure Plans that would be needed under legislation then passing through Parliament, and for investment and other major decisions by central government (ibid.)

The resultant *Strategic Plan* appeared in June 1970 together with five bulky volumes of supporting evidence. In every way it was the logical successor to the Abercrombie Plan of a quarter of a century before, providing a new set of strategic policies for the long-term development of the region. It accepted that both population and economic activity would continue to expand in certain parts of the Outer Metropolitan Area, above all in Reading–Wokingham–Aldershot–Basingstoke, which could be expected to grow by no less than 200 000, or 37%, in the brief period from 1966 to 1981.

The *Strategic Plan* argued that this growth should be concentrated into a few major growth centres – each a polycentric bundle of large old towns, new and expanded towns, smaller market towns and villages, separated by open green land – among which the Reading subregion was an outstanding example. It concluded thus from an evaluation of two alternative strategies: 1991A based on distant counter-magnets 50 miles (80 km) and more from the capital, 1991B based on nearer-in concentrations 30–40 miles (50–60 km) away (GB South East Joint Planning Team 1970, pp. 56–64). Both, significantly, gave an important role to the Reading subregion. In fact the two hypotheses did not differ nearly as much as they appeared: the real issue was not the continuation of containment-overspill policies, but the kind of overspill, the scale of the proposals, and above all the relative role of the public sector, greater in A than in B.

The result was a classic British compromise: a recommended strategy combining parts of both. A limited number of major growth centres would be developed at varying distances from London, some in the OMA, some near the region's fringes. South Hampshire, 70–80 miles (110–30 km) distant, should grow to 1.4 million by the century's end; Milton Keynes–Northampton–Wellingborough, 50–70 miles (110–30 km), to 0.8 million; South Essex, 30–45 miles (50–70 km), to 1.0 million; Crawley–Gatwick, 25–35 miles (40–60 km), to 0.5 million; and finally Reading–Wokingham–Aldershot–Basingstoke, 30–50 miles (50–80 km) from London, to between 1.0 and 1.2 million (GB South East Joint Planning Team 1970, pp. 81–2).

The *Strategic Plan* had a distinct and deliberate western bias. Three of the five proposed major growth areas were located in the western hemisphere of the OMA – Milton Keynes, Reading–Wokingham–Aldershot–Basingstoke, and South Hampshire – and of the remaining two the only eastern growth area proposed in the *Plan*, South Essex, was subsequently dropped by the government.

The intention, clearly, was to open up the region by establishing city-regions that were both self-contained and mutually accessible – so that, in the Reading area, towns and villages would functionally coalesce into a polycentric city region that could easily provide for most needs of its inhabitants, whether for employment, shopping or entertainment (Hall *et al.* 1973, p. 478). In 1971 central government gave the *Strategic Plan* its blessing. Around Reading, as elsewhere, the local authorities also accepted the principles and the overall recommendations as a basis for structure plan preparation and national investment; but they reserved their judgement on details of scale, timing and implementation (Study Team 1975, p. 2). This was small wonder, because a major reorganization of local government was slowly and contentiously passing through Parliament.

The Reading subregional study

Area 8 – the Reading–Wokingham–Aldershot–Basingstoke area, identified in the *Strategic Plan* as one of the five major growth zones – straddled the boundaries of Berkshire, Hampshire and Surrey, which were busy preparing separate structure plans. In January 1973, a joint planning team and steering committee were created to co-ordinate this work – and, for obvious reasons, their remit was extended to cover the whole of the structure plan areas of central Berkshire and north-east Hampshire. The team's report appeared in September 1975.

This, as the 1964 *Study* had predicted, was truly a zone of growth. From 1951 to 1961 population had increased by 85 000; from 1961 to 1971 by another 174 000, to a total of 623 000. But latterly, falling birth rates – and diversion of migration streams out of the South-East altogether – had caused the growth rate to plummet, to a mere 23 000 from 1971 to 1973 (Study Team 1975, p. 12). So future predictions must be revised sharply downwards: against the *Strategic Plan's* 2.8 million for the entire South-East from 1971 to 1991, a mere 600 000. Nevertheless, a dynamic area like this would continue to attract new people, bringing problems for the rest of the region and indeed for this area itself (ibid., p. 199).

Employment too had grown rapidly – to 255 000 in 1971 – and unemployment was a minimal 1% during the 1960s; the chief problem, indeed, was severe labour shortage (ibid., p. 13). The decline of traditional sectors had been more than offset by growth in new ones: while agricultural employment had fallen some 30%, electrical engineering jobs had increased three and a half times in the previous decade. And professional and scientific employment had also burgeoned, at least partly through public sector research jobs: the GREs alone employed an estimated 15 000 in 1971; the armed forces – particularly the Army – employed another 15 000 directly, 6000 indirectly (ibid., p. 13).

Reviewing these trends and evaluating strategic options, the team noticed that much of the predicted growth would be housed through existing plan commitments around Reading itself, and in new and expanded towns. Bracknell new town would grow from a mere 5100 in 1951 to 60 000 in 1984; the expansion of Basingstoke would take it from 17 000 in 1951 to 113 000 in 1986 (ibid., p. 12). These commitments gave 68 500 new dwellings: 15 800 in Greater Reading, 11 300 in Basingstoke, 4000 in Frimley and Camberley, the rest in settlements throughout the region. Everywhere, growth would be constrained by the need for new infrastructure and the extensive landholdings of the Ministry of Defence. All this suggested a policy of concentration (ibid., p. 217).

The sharply reduced population projections made this feasible. A year after the study team's report, an official review of the *Strategic Plan* confirmed this last point: against the 1970 figure for the whole South-East of an additional 3.7 million between 1971 and 1991, the new projection was a mere 1 million, not quite as drastic as the study team's assumption of 600 000, but still dramatic (GB South East Joint Planning Team 1976, para. 1.8). And, the review underlined, public sector investment would be tighter in the 1980s than in the 1970s: a prediction that proved only too true.

The structure plans and after

The structure plans for Area 8, produced in the immediate wake of the Study Team's 1975 report – central and east Berkshire in 1978, north-east Hampshire and Surrey in 1980 – all reflected this new pessimism. True, the central Berkshire planners noted, housing completions continued high and houses sold quickly; despite continued job losses in traditional sectors, high technology industries were creating new employment. They proposed to reconcile the demands for growth and the desire for conservation in a traditional way: by concentrating new development in existing settlements. Some 4066 acres (1484 hectares) of housing land – part with existing permissions, part not yet – was provided, notably at Tilehurst, west of Reading, and Earley, south-east of it, west of Wokingham and south-east of Bracknell; there was a promise to release more land, not then specified, north-east of Bracknell; otherwise, there were to be no further major residential land releases. This policy bundle was to meet a projected population growth, between 1976 and 1991, ranging from 58 800 (16%) on the most optimistic assessment to as little as 21 700 (6%) on the lowest; clearly, the county would seek to keep growth down to the lower end of this range (Royal County of Berkshire 1978, pp. 36–43). Major allocations for industrial and warehouse growth were made close by: south of Reading at Manor Farm and Worton Grange close to the M4/A33 interchange, south-east of Reading at Winnersh Triangle, in the southern industrial area of Bracknell (ibid., pp. 47–8). Such policies would be balanced by severe restraints on development elsewhere, especially in the green belt east of Wokingham, and the County Council would seek to maintain 'a clear physical and visual distinction between built areas and the countryside' (ibid., p. 79).

In north-east Hampshire the same policy of concentration was evident: above all in Basingstoke, which had half the housing provision, and in Aldershot–Farnborough–Fleet which had another quarter (County of Hampshire 1978, p. 36); here also, in Basingstoke, Farnborough and Fleet, new industrial development would be concentrated (ibid., p. 29). Again, housing development in the open countryside would be strictly controlled and 'The loss of, or disturbance to, agricultural, horticultural and forestry land will be minimised' (ibid., pp. 39, 75).

Local planners and politicians were therefore making their policies clear: new employment and new housing, a necessary evil, would be grouped where they could do least harm to the area's natural environment. In general, central government was willing to accept this compromise; but not without disagreement, sometimes major. In accepting the Central Berkshire Structure Plan, Michael Heseltine as Secretary of State for the Environment modified it to add provision for 8000 additional houses; dubbed Heseltown, the proposal brought acrimonious reaction from local councils and local residents, as a hapless county sought one way or another to find the necessary land. In Spring 1984 the county's revised structure plan proposals placed half of Heseltown on the site north-east of Bracknell, staked out for just such a purpose in the 1978 version; renewed uproar was the inevitable result.

Conclusion: the role of the public sector

Though it is sometimes suggested that development of the Crescent is a shameless triumph of privately led growth, our contention is that the public sector has played a crucial role both in paving the way for this growth, and providing much of the capital to generate and sustain it. Specifically, this role has been fivefold.

First, conventional land use planning has had a critical effect – but not always the one conventionally assumed by practitioners or protagonists. Though 'positive planning' has been significant – a new town at Bracknell, major town expansions of Basingstoke and Andover, smaller ones elsewhere – the major contribution has been 'negative planning': the attempt to *prevent growth taking place*, particularly in places where it is thought environmentally harmful. This aim of environmental protection has been at least as significant an objective of policy as has economic development. Indeed, until recently, economic growth and full employment were taken for granted here.

Yet, paradoxically, negative planning has proved instrumental in making the area attractive to high tech companies and research organizations. Such activities are fairly free in their locational choice; they are influenced by the preferences of their skilled scientific, engineering and technological employees for a good living and working environment. Indeed, as Chapter 5 has shown, the residential choices of owners and managers have often proved crucial in the location of an infant firm. Thus, in unconsciously following Cromwell's approach to government – 'more surely what they wouldn't have than what they would' – local planners have actually created the conditions for high tech growth.

Secondly, and relatedly, a particular *type* of planning – large-scale, long-term, strategic – has been particularly influential, for a generation and more, in mediating the continued trade-off between physical expansion and environmental protection. Though currently every-where unfashionable, its value has lain in its ability to provide long-term continuity for the protection of the countryside in an area under severe pressure from inward investment.

Thirdly, where growth has been permitted, it has proved remarkably successful – at least, if success is measured by conventional measures like relatively low unemployment and high incomes. Yet the key regional development agencies were not the conventional ones. They were the Ministry of Defence, the Civil Aviation Authority, the former Board of Trade, the Departments of Transport and Education and Science. These public sector bodies have created tens of thousands of new jobs directly and indirectly: through the construction of government research establishments, the carrying of cargo by air, the granting of industrial development certificates, the building of high-speed roads, and the award to local firms of research and development contracts.

Fourthly, in all this activity there has been remarkably little in the way of conscious strategic planning with the specific aim of creating an area of high tech growth. The Western Crescent happened largely by accident, as a series of unco-ordinated sectoral decisions made by public agencies, which combined spatially to produce an attractive environment for modern industries. Though there were notable strategic planning exercises (in 1943–4 and 1964–70) they never recognized the specific objective of building up a major area of high tech based industry west of London. The various sectoral agencies, particularly those engaged in defence, were simply – though often grudgingly – allowed to go their own way.

Fifthly, over the past four decades there has been a steady shift in the orientation of central government and thus in its relation with local authorities in the Crescent. The environmental protection stance, maintained for decades, has increasingly given way to a much shorter-term, economic-oriented, 'invest in success' approach. This has two elements: first, the belief that national economic needs are more important than local amenity, and that growth – particular technologically based growth – should be encouraged, even if local concerns have to be over-ridden; secondly, that much of the development required, particularly the large-scale provision of housing, can be provided almost exclusively by the private sector. Though the results of this approach have become apparent in the 1980s, its origins go back to the 1960s. Whitehall has increasingly assumed the role of high tech urban economic growth promoter; in consequence, the Crescent's local authorities have been left fighting a desperate rearguard action to protect their rural environments, on behalf of the existing residents.

11

The infrastructure base

Public decision makers did more than control the area's land uses. By providing key pieces of infrastructure, they could promote its development – and, conversely, by failing to do so, impede it. In practice, some of this infrastructure was so local, so taken for granted, that it could have little effect. That particularly applies to the more ubiquitous local facilities such as water, electricity or telephone service. Even in 1951, the survey for the original Berkshire county development plan could report that there were very few settlements in the whole county where electricity was not available (Berkshire 1951, p. 65); water supplies, similarly, met demands in the more closely settled parts, though mains did not cover the whole area even in such areas as Wokingham Rural District (ibid., p. 67); sewers were still to be provided in many places, even those close to the towns such as Woodley and Earley east of Reading or the area between Maidenhead and Twyford (ibid., pp. 71–2). But these deficiencies could hardly have proved a major handicap, particularly since most were remedied in the next few years. Similarly, in so far as telephone waiting lists continued to frustrate potential customers throughout the 1950s and 1960s, the position in the Corridor was little better or worse than most other places.

Far more important, potentially, were the decisions that created major national pieces of transport infrastructure, which provided crucial linkages to London, to other parts of Britain and to the world. By the late 1970s, these decisions had already given the Western Crescent in general, and the Western Corridor in particular, an outstanding nodal position in the complex network of Britain's national and international spatial linkages. We have seen, in Chapter 5, that many firms in the Corridor specifically mentioned these linkages as factors in their own location.

It is important to make the point that the area stretching west from London to Bristol is a natural transport corridor, and has been recognized as such for centuries. The area is relatively open and flat with very little in the way of major physical obstacles to movement. It was no accident that Brunel chose the area to build his Victorian high-speed train network, and his work has been followed by a great deal of subsequent transport investment.

But how far were these infrastructure investment decisions part of a conscious plan? How far did decision makers at national level decide that the Corridor was to be invested with these unique advantages? Other decision makers, during most of the period from 1945 onwards, were working in a quite opposite direction: to discourage private industrial investment in this area and to divert it, if possible, to the Assisted Areas of northern and western Britain. To that end, they sought *inter alia* to improve motorway and trunk road facilities into and within the Assisted Areas, to improve regional airports, and otherwise to bring the peripheral regions of Britain closer to the South-East core. But, it appears, at the same time – and even, perversely, sometimes as part of the policy – they enhanced the accessibility of the Corridor.

The task of this chapter is to trace precisely how and why this quirk of history occurred. Three major pieces of transport infrastructure will provide case studies: Heathrow Airport, the M4 motorway and the Inter-City 125 trains.

Heathrow

The origins of Heathrow airport are still shrouded in a cloak of wartime secrecy. A site nearby, on Hounslow Heath, had been the scene of the first international civil aviation services in Britain, in 1919 (Allen 1983, p. 75). The Heathrow site had been earmarked in 1928 by Fairey Aviation as a small general aviation field, and had opened in 1930 as the Great West Aerodrome (Wright 1983, p. 33). Then, sometime in 1943, it was taken over by the newly formed Transport Command of the Royal Air Force to be developed as an RAF aerodrome capable of taking the heaviest wartime aircraft, and in particular for trooping movements to the Far East. Apparently, however, in examining some 50 possible sites, Transport Command took care to find a site suitable for postwar development as London's first airport site (GB Ministry of Civil Aviation 1948, p. 6; Allen 1983, p. 75). In fact, because of this aspect, funds were not immediately forthcoming and construction began only in 1944 (Wright 1983, p. 33).

In thus preparing for peace, Transport Command was acknowledging the strength of a vocal lobby which had long campaigned in and out of Parliament for the creation of a top-quality international airport for London. That this was lacking, no one could doubt. Both Croydon and Heston, London's prewar airports, lacked even a proper hard runway. The reason, MPs were arguing, was the dead hand of the Air Ministry, which has, as Captain Macdonald put it in a debate on 3 August 1944, 'clung to this service for all the years before the war, with the result that the history of British civil aviation has been one of frustration, procrastination and ineptitude' (Commons Hansard 1943–4, **402**, 1699).

Throughout the second half of World War II, and even beyond its end, MPs and peers repeatedly pressed the government for a statement on the new London airport. Equally resolutely, ministers resisted their pressures. The resulting exchanges finally reached the level of farce, in which a cherished state secret had become public knowledge. The long saga began on 15 December 1943, when, opening a debate in the Lords, Viscount Rothermere declared: 'I feel that until some decision is come to upon a London airport, the success of an air transport system after the war cannot in any circumstances be certain' (Lords Hansard 1943–4, **130**, 341). The new airport, he pointed out perceptively, could become 'the junction for Europe' (ibid., p. 343). Evidently, he was concerned at rumours that the government might try to establish the major airport hub outside London, perhaps at Prestwick or at Hurn near Bournemouth (ibid., p. 344): 'It is essential that passengers arriving in this country should be landed at an aerodrome within, say, fifteen miles of London, so that they can be taken to their accommodation in the centre of London by motor car' (ibid., p. 342). The selected site, he went on to argue, could be used for war purposes; 'and this would be one of the few expenditures of money for war purposes which would give us a certain asset after the war' (ibid.). Lord Sherwood, in replying for the government, was Delphic:

> in planning our war-time airfields, especially those near large centres of population, we have had in mind, so far as military considerations allow, the needs of peace, and a chain of airfields has been provided which should go far to satisfy the requirements of post-war civil transport. (ibid., pp. 361–2)

In fact, as subsequently emerged, the choice of Heathrow had already been made – and on precisely the criteria that Rothermere had suggested. Here, the role of RAF Transport Command was probably crucial. Its birth on 25 March 1943, to co-ordinate all military transport (Saunders 1954, pp. 184–5), had provoked a huge row with the directors of the British Overseas Airways Corporation, who felt that their role in providing scheduled air services had been usurped; four of the five Board members, including the Chairman, had resigned (GB Air Ministry 1943, pp. 5–6). Hence, in subsequent negotiations the Air Ministry presumably felt that it had to be particularly careful about the views of the new Board. Subsequently, the Parliamentary Under-Secretary for Civil Aviation confirmed that there was close consultation between the Air Ministry and his department to ensure that work for military purposes could be incorporated into the civil development scheme (Commons Hansard 1945–6, **414**, 212).

But this admission had to wait until 10 October 1945. During 1944 and early 1945, questions from MPs showed clearly that they knew Heathrow was under construction but the government was refusing to say so. Twice, on 22 March and 6 July 1944, Captain Balfour – the Secretary of State for Air – declined to give anything away. On 3 August, Mr Bowles, MP for Nuneaton, returned to the fray:

> It seems that the Government must have some plan, because I have recently seen a most marvellous aerodrome in course of erection, with runways five miles long [*sic*]. Messrs. Wimpey have the contract, and it must be one of great cost to the country. So there is a policy being worked out administratively, and the House of Commons is being kept in the dark. (Commons Hansard 1943–4, **402**, 1699)

Captain Balfour was determined as usual to give as little as possible away:

> let me assure my hon. Friend that that aerodrome is being made for military purposes – for military transport purposes. We see great new military transports coming along and we need a new airport near to the heart of Government. As and when, after the war, we decide what airports are to be used for civil purposes, all suitable airports and their possibilities will be surveyed and the best one chosen . . . it is not being done by any civil authority, but under military authority, as it is to be a military airport for the war against Germany and against Japan, which involves long communications. (ibid., 1700, 1710–11)

On 6 December 1944 Captain Balfour's successor, Sir Andrew Sinclair, repeated the usual formula, provoking a wry exchange:

> *Mr Bossom*: Has my right hon. Friend indicated that there is not going to be a new airport at Staines?
> *Sir A. Sinclair*: No, Sir. The airfield being constructed there is being constructed for military purposes and is urgently required.
> *Mr Bowles*: May I ask the right hon. Gentleman whether he does not think that the construction of the airport may not be completed before the war is over?
> *Sir A. Sinclair*: I think we are almost certain to bring that airport into use before the end of this war. (Commons Hansard 1944–5, **406**, 512–3)

Mr Bowles, it transpired, was right and the Secretary of State wrong. Meanwhile, on 26 January 1945 Sir Stafford Cripps again denied that a decision had been taken (Commons Hansard 1944–5, **407**, 1252–3). Doggedly, on 6 March Mr Montague referred in his question to 'the Middlesex airport called Heath Row, the new London Airport that is being built' (ibid., **408**, 1866). By this time, also, Abercrombie's *Greater London plan*, referred to in the previous chapter, was being published. He was in no doubt: Heathrow was to be London's chief airport, for 'in fixing a site which will obviously be of extreme importance it is desirable to secure ample room' (Abercrombie 1945, p. 79). The site, 3.5 by 3 miles (5.6 by 4.9 km) in extent, was on level land with a gravel subsoil:

> Although the site of the whole airport is on land of first-rate agricultural quality, it is felt after careful consideration and thorough weighing up of all the factors, that the sacrifice for the proposed purpose of the part south of Bath Road is justified . . . Surface communications will be provided to London: (a) by road, *via* express arterials 2 and 1; (b) by electric railway to Waterloo and Victoria by means of a short branch to the Southern Railway line west of Feltham. A two-mile extension of the tube railway from West Hounslow would also give direct communication with the Underground system. (ibid., p. 80)

So intense was the interest by this time, that the government evidently felt the need to issue a retraction. On 22 February, under the heading 'No decision on conversion of Heath Row', *The Times* Aeronautical Correspondent emphasized that

> As has been stated more than once in Parliament, Heath Row is being built for military purposes. It was started at a time when victory over Germany seemed remote, but now, when the airfield is still far from complete, its use for military purposes seems doubtful . . . An early decision on the future of Heath Row is desirable. (*The Times*, 22 February 1945, p. 12)

It would, he believed, make an admirable landplane airport for London; though it lacked water facilities, it might be possible to construct a flying-boat base nearby (ibid.). This might suggest that the near presence of old gravel pits may have been one of the factors in Heathrow's selection.

On 15 March Mr Granville reopened the game in the Commons by referring to the 'land known as Heath Row and its environs', and asking 'when this airport will be finished to take its place as World Number One Air Terminus'. But as ever, the government spokesman refused to rise to the bait (Commons Hansard 1944–5, **409**, 468). Indeed, as late as October 1945 – five months after VE Day, and fully two months after the end of hostilities – Mr Thomas, the Parliamentary Secretary for the Ministry of Civil Aviation in the new Labour government, was still insisting that the work was on behalf of the Air Ministry and that the *long-term* policy was for its civilian use. And in

October, a month after completion of the main runway, RAF Transport Command actually took over. Not until December did the government produce a White Paper which, as well as establishing BEA and BOAC as separate nationalized corporations, declared at last that

> Heathrow will be designated as the long-distance airport for London and will be developed to the highest international standards required for trans-oceanic aircraft. Croydon and Northolt will be used as London airports for internal and European traffic in the period immediately ahead. (GB Minister of Civil Aviation 1945, para. 21)

Shortly after this, on 24 December, the Ministry of Civil Aviation could finally announce that it was to assume the administration of 'Heathrow, the designated airport for London' (*The Times*, 24 December 1945, p. 2). This was just over two weeks after the first test landing, and one week before the first scheduled flight took off on 1 January 1946 (GB Minister of Civil Aviation 1948, pp. 7–8; Wright 1983, p. 33). The airport – renamed London Airport, because Whitehall had learned that foreigners could not pronounce Heathrow – was officially opened on 25 March 1946 (McKie 1973, p. 30).

This extraordinary saga of prevarication evidently resulted from bitter internecine strife between the Air Ministry and the new Ministry of Civil Aviation, which had finally been set up in October 1944 to take responsibility for this area out of the Air Ministry's hands (ibid., p. 30). It had at least one unfortunate permanent result. The RAF's invariable policy was to design all its airports on a triangular pattern of runways, evidently on the basis that this gave greater flexibility in combat conditions (Allward 1966, p. 4). In September 1945, even before the Civil Aviation Ministry took over, it appointed an Advisory Panel to produce a plan for Heathrow's long-term development. The resulting report, published in May 1946, came out in favour of staggered runways, with one major runway paralleled by two shorter ones. But it also stressed the need to make the best use of the triangle of runways already open or under construction.

The result was a very complex design of no less than nine runways arranged in three triangles: the first, consisting of the RAF inheritance, constituting Stage One of development; the second, superimposed upside-down on the first to create a Star of David pattern, constituting Stage Two; and the third, forming a separate triangle north of the A4 road, making an eventual Stage Three. In the event, Stage Three was abandoned; and three of the remaining six runways were eventually retired. The resulting pattern today consists of two long parallel runways constituting the bases of the first two triangles, which are aligned almost due east–west, plus one of the other sides of the first triangle aligned diagonally to them; this last is used only for landing in the unusual event of a southwesterly gale, so most of the point of the original design has been lost.

But this design, and in particular the RAF inheritance, have had two enduring and most unfortunate consequences. The first is that all terminal facilities were locked into a 60-acre (148-hectare) central island site, a pattern broken only in Spring 1986 by the opening of Terminal Four on the airport's southern perimeter. Strangely, the forecasts of the 1946 report – 100 aircraft an hour – proved quite robust; what the panel failed to foresee was the jumbo jet and the resulting growth in groundside traffic which the island site is singularly ill-suited to handle.

The second consequence arose from the location due west of central London, coupled with the alignment of the two main runways – which would have been necessary, RAF or no RAF, because of the prevailing wind. (Gatwick, the two Paris airports, Amsterdam and Frankfurt all use similar alignments.) It meant that in the most usual, cyclonic, weather conditions the approach path would lie right over intensely built-up areas. No one, it seems – not even Abercrombie – appreciated the problem of jet noise, though by 1945 the first military jets had been flying overhead for over a year. Lord Swinton, the first Minister for Civil Aviation, declared: 'I say without the least hesitation that this was the only possible site on which a great airport for London could be built' (McKie 1973, p. 30). And his successor, Lord Winster, added: 'One of the reasons which led to the siting of the London airport in this district was that it could be done with the minimum disturbance of householders' (Lords Hansard 1945–6, **140**, 702).

He did not prove highly prophetic. In this sense, Heathrow did not come to be the jewel in the international aviation crown that the planners of 1946 imagined. But, despite its limitations, it achieved the essential target to which the Attlee government aspired: thanks partly to an early start while most of Europe's airports lay ruined, it soon took a commanding lead as Europe's first airport and the world's leading international airport. And this was further reinforced in 1953, when a further White Paper recommended that the bulk of all remaining flights should be diverted from Northolt – which henceforth should be restricted to military use – into Heathrow (GB Minister of Civil Aviation 1953, para. 3).

Thus London's major international airport came to be located almost exactly due west of its centre: a location that in the jet age, beginning just over a decade later, was to pose a major environmental problem. But London's problem was Berkshire's opportunity.

The M4

The second major piece of national infrastructure in the Western Corridor, in contrast, was planned almost in a cacophony of publicity. Its story, too, begins before World War II. At that time, as until the beginning of the 1960s, the route to the west was the old Bath Road with the addition of bypasses built as part of the Arterial Roads programme of the 1920s: the Great West Road around Brentford and Hounslow, which had first seen light in a report of the Board of Trade Traffic Branch in 1910, a further bypass of Cranford, and a bypass of Twyford. The line had been fixed for bypasses of Slough and Maidenhead, and work on the Maidenhead section had started in 1938, to be summarily stopped on the outbreak of war.

In May 1938 the County Surveyors' Society produced a national plan for 1000 miles (1600 km) of new motorway, which included *inter alia* a line from London via Reading to Bristol, crossing the Severn by a new bridge and so to Cardiff and Swansea (Drake *et al.* 1969, pp. 36–8). The Minister of Transport, after cogitation, was duly cautious. In a statement of May 1939, he commented:

> The German motor roads have excited the admiration – and excited the imagination – of all who have seen them. There can be no doubt that they afford first-rate opportunities for long-distance motor traffic to move fast and with a great measure of safety. Their potential advantages are so obvious as hardly to need statement. (ibid., pp. 38–9)

Nevertheless, in applying the principle to Britain there were problems. The cost of construction would be greater. The reduction in accidents could be smaller because most occurred in towns. Limiting roads to motor traffic would be a novelty requiring legislation. Soon after that, the war intervened (ibid., pp. 38–41).

Before its end, though, the government had announced a change of mind: it was preparing plans for motorways. Abercrombie, in drawing up his *Greater London plan* of 1944, evidently felt sufficiently confident about their prospects to use them as a major element of his proposed regional structure: motorways to South Wales and to Exeter would radiate outwards from a series of orbital highways running through London's built-up fabric – or, in the outermost case, through the green belt that was to limit the conurbation's growth:

> Acting upon the implications of the classification of the Ministry of Transport, the class of Trunk Roads has been transformed into a system of express arterial highways which are intended to canalise the through traffic: these link up with the ten main radials of the County of London Plan, on the one hand, and are extended beyond the Region in a series of national routes. These radials are directed inwards to one ring within the County of London and traverse another situated outside the suburban built-up area. (Abercrombie 1945, p. 66)

Specifically, the plan contained an M3 almost on its future line, and an M4 starting from a more southerly origin south of Heathrow but then following the line that would be completed more than a quarter of a century later. Both these linked in to the outermost orbital or D Ring, at the edge of the conurbation. In addition, there was to be an orbital parkway built through the green belt a little further out. The actual M25 was to follow approximately this outer line, but in places – especially north of London – it would swing inwards, to pick up the route of the D Ring.

In May 1946 the then minister, Alfred Barnes, exhibited an 800-mile (1288 km) plan in the House of Commons tea room (Starkie 1982, p. 2). It was to all intents and purposes the programme as eventually completed between 1959 and 1972, and it included a London–South Wales motorway which now, however, stopped short of Cardiff (Drake *et al.* 1969, p. 44). Soon after this, the ministry asked the Berkshire County Surveyor to investigate a possible line running south of Reading and Newbury (Gregory 1967, p. 119). In 1949 the Special Roads Act at last provided powers for the construction of roads restricted to motor traffic.

Then, spending cuts intervened. Ironically, the main reason was the diversion of government priorities into defence, which provided a major boost to the industries of the Corridor while delaying a start on its highway infrastructure. But eventually, on 2 February 1955, the Minister of Transport, John Boyd Carpenter, announced the first instalment of an expanded roads programme, including the Maiden-head By-Pass (Commons Hansard 1954–5, **536**, 1109). Then, in a statement of 22 July 1957, his successor Harold Watkinson committed himself to five major priority schemes including 'the road to the west', to end either at Bristol or the Severn Bridge (Commons Hansard 1956–7, **574**, 52, 57; Starkie 1982, p. 5). Finally, on 30 July 1959, Mr Watkinson went further, promising 'Construction of the South Wales Radial Road from London to South Wales via the Severn Bridge' as part of the 'creation of a network of national trunk roads' which would 'give the country a basic road system on which traffic will flow smoothly and safely'. Priorities, the minister announced, would be 'settled according to traffic needs, particularly those of industry and commerce' (Commons Hansard 1958–9, **610**, 161).

One small part of this plan gave no trouble at all. Eleven miles of it, by-passing Slough and Maidenhead, followed a prewar line already incorporated in the Berkshire and Buckinghamshire Development Plans. Indeed the Maidenhead By-Pass, first proposed in the 1920s, had been started before World War II and the earthworks had been lying derelict ever since. So the scheme made in February 1957 fixed this first section and it was opened, in sections, by the early 1960s. The 12-mile (19 km) stretch from Slough into London was more controversial because of its plan for a viaduct over the existing A4; but it too was fixed in October 1960 and opened in 1965 (Gregory 1967, p. 119). Meanwhile, at the other end a stretch from Almondsbury to Tormarton on the A46, bypassing Bristol, was fixed in September 1961. Thus, at this date, there was a gap between 70 and 80 miles (113–129 km) long (ibid., pp. 119–20).

The most obvious way to fill it was the Bath Road route following the old A4, inherited from the 1938 plan and running south of Reading and Newbury. In fact the first 13 miles of this route, from Maidenhead to Three Mile Cross south of Reading, had actually been incorporated in the 1958 review of the Berkshire Development Plan (ibid., p. 120). At this point, however, the Ministry of Transport began to have doubts: the route was not the shortest, hence not the cheapest, way from Maidenhead to Tormarton; and it failed to serve Swindon, recently designated as an Expanding Town under the 1952 Town Development Act. So in 1961, after commissioning a Consultant's report, the ministry swung in favour of a 'Direct Route', running north of Reading and on to Liddington near Swindon.

Thereafter, for four years, a battle raged. In the first place, the Ministry of Housing and Local Government – precursor of today's Department of the Environment – continued to favour the Bath Road route, at least as far as a point south of Newbury, at which point they wished the motorway to swing north-west to serve Swindon. Secondly, both Oxford and Reading were vehemently opposed to the Direct Route – Oxfordshire because it invaded its southern Chiltern slopes, Reading because it would generate too much cross-town traffic. Thirdly, equally weighty bodies like the National Parks Commission and the Royal Fine Art Commission were opposed. And finally, so were big amenity groups in the area of the Berkshire Downs. A desperate attempt by the ministry to mollify this opposition, through a proposal for a Revised Direct Route a little to the south of the original line, failed completely in its objective. Yet, to the south, the weighty Kennet Valley Preservation Society were equally opposed to the Bath Road route (ibid., pp. 126–7, 271–5).

Finally, in February 1964 three new factors together swung the ministry into a major change of mind. First, a Reading Traffic Survey confirmed that 90% of traffic in the town had origins and destinations south of the main east–west railway line. Secondly, the Ministry of Housing and Local Government's *South East study* recommended a major new city at Newbury. And thirdly, the Ministry of Transport became convinced that eventually, because of future overloading on the M4, a new branch would be needed, running into south-west London. All these pointed in the same direction: to a line south of Reading, following the Bath Road route.

That would offend the preservers of the Kennet Valley. But, in January 1964, they had already provided the answer in the form of a consultant's report suggesting that indeed the M4 should run south of Reading, but should then swing north-west to follow approximately the line of the Revised Direct Route running just north of Newbury and thence across the downs to Swindon. The ministry officials seized upon this way out of the morass: it met almost all the serious objections, it offended almost no one badly, and although many bodies were not entirely happy with it they were willing to live with it (ibid., pp. 282–3). Gregory, in his study of the line of the M4, concludes:

> Had the nature of the countryside between Maidenhead and
> Liddington been different, and had the districts though which the
> various lines passed contained fewer persistent, determined and
> articulate individuals, an earlier – and therefore less satisfac-
> tory – planning decision might well have been reached. (ibid.,
> p. 286)

Certainly, from a 20-year historical hindsight, the final decision was the
best, or the least bad, one. High tech industry was already developing
south of Reading and in the corridor eastwards from there to Bracknell,
which the chosen line of the M4 bisected. The motorway did not create
this growth, which was already taking place without it. But it
admirably served it, and thus helped promote it. The direct line would
have done no such thing, but would indeed have created impossible
traffic problems within Reading as well as pressures for development
in the sensitive Thames Valley landscapes on its northern side. As
Gregory suggests, the long delay in fixing the line proved a blessing in
disguise.

It took another seven years to go through the complex statutory
procedures and finally to build the road. On 22 December 1971 the
Minister of Transport, Michael Heseltine, cut the ribbon at Holyport
near Maidenhead and got into his car to lead the procession down the
50-mile (80 km) piece of new motorway to Badbury in Wiltshire – one
of the longest single stretches ever to open at one time in Britain. Farce
intervened when the minister's British-built luxury car refused to start.
To the very last, the M4 had resisted coming into existence. And by
this time, ironically, the M4 Corridor was already an evident fact.

Inter-City 125s

In one important sense, the newest piece of transport infrastructure in
the Corridor is not new at all. British Rail's Inter-City 125 service – the
first anywhere on the BR network – was inaugurated between
Paddington, Bristol and South Wales only in October 1976, 31 years
after the opening of Heathrow, and six years after completion of the
M4. But the roadbed on which it ran was built between 1835 and 1841.
And what made it possible – indeed, what made this the obvious
choice for the launch of the new trains – was the exceptional standard
to which this road had been engineered.

The explanation lay in the extraordinary personality of its engineer.
Isambard Kingdom Brunel, who designed and supervised the construction of the line between the ages of 27 and 33, was in no doubt that he
was designing it for posterity when he set out to build 'the finest work
in England' (Rolt 1970, p. 188), and he did. The historians of the Great
Western Railway comment: 'No other trunk line, perhaps in the world,

has required less basic alteration for increasing speeds than the original GW out of Paddington' (Whitehouse & Thomas 1984, p. 18).

The Japanese Shinkansen and the French TGV both required huge expenditures on a totally new railway built to new standards of alignment. The first British high speed trains merely required resignalling and some minor track works. It was almost as if Brunel had built the line for them – as, in a sense, he did. Justifying his choice of the broad gauge, he told the Gauge Commissioners in 1845:

> Looking to the speeds which I contemplated would be adopted on railways and the masses to be moved, it seemed to me that the whole machine was too small for the work to be done, and that it required that the parts should be on a scale more commensurate with the mass and velocity to be attained. (Rolt 1970, p. 147)

Certainly, on Brunel's main line they were. The broad gauge has gone, a victim of the battle he lost, but the rest of the design remains. For 71 of the 118 miles (114/190 km) from London to Bristol, the line has a maximum gradient of 1 in 1000; in other words, it is effectively level. For the first 53 miles (85 km) from Paddington to Didcot, the steepest gradient is 1 in 1320; for the next 24 miles (39 km) to the summit at Swindon, the steepest are two short stretches at 1 in 660; from Swindon to the crossing of the Avon at Christian Melford, the steepest is also 1 in 660. Thus virtually all the steep gradients, 1 in 100, are concentrated into a mere four miles (6.4 km) immediately east of Bath and in the Wootton Bassett cutting (ibid., p. 185). To achieve this feat, Brunel took the line all the way up the Thames almost to its headwaters, thence crossing a low interfluve to the upper Avon before cutting boldly through the obstacle of the Cotswolds via the Box Tunnel, which his many opponents thought an impossible and dangerous thing to do. Where he encountered obstacles, as in the low plateau at Sonning immediately to the east of Reading, he cut ruthlessly through them to maintain the even quality of his road: the Sonning Cutting, 2 miles (3.2 km) long and 60 feet (18.3 m) deep, required 1220 navvies and 196 horses moving 24 500 cubic yards (18 743 m^3) spoil a week (ibid., p. 174).

Brunel was quite consciously designing his railway for extraordinary speed, as he made clear in a statement in August 1838:

> The peculiarity of the circumstances of this railway . . . consists in the unusually favourable gradients and curves, which we have been able to obtain. With the capability of carrying the line upwards of 50 miles out of London on almost a dead level, and without any objectionable curve, and having beyond this, and for the whole distance to Bristol, excellent gradients, it was thought that unusually high speeds might easily be obtained, and that the very large extent of passenger traffic which such a line would undoubtedly command, would ensure a return for any advant-

ages that could be offered to the public, either in increased speed
or increased accommodation. (Quoted in MacDermot 1927, vol. I,
p. 65)

Even in the early days, therefore, the Great Western was able to
achieve speeds that were extraordinary by the general standards of the
time. Between 1847 and 1852, the Flying Dutchman travelled the 53
miles (85 km) from Paddington to Didcot in 55 minutes at an average
speed of 58 miles an hour (93 km/hr) (Whitehouse & Thomas 1984,
pp. 31–4). Thence, after Brunel's untimely death at the age of 53,
standards slipped. Not until 1929 did the Great Western again exploit
its inheritance, when the Cheltenham Flyer achieved 66 miles an hour
(106 km/hr) between Swindon and Paddington – later increased to a
dizzying 71.4 mph (114.9 km/hr). But such an operation was difficult
and even embarrassing to maintain: the braking systems of the time
were not good enough, so that special signalling arrangements were
necessary to keep seven blocks open at a time instead of the normal
one (Nock 1980, pp. 10–11). Because of this, the Cheltenham Flyer –
like the Flying Dutchman before it and like other famous expresses of
the 1970s – was a prestigious freak. The great majority of Great
Western trains trundled along at much more sedate speeds (ibid.,
p. 22).

In fact it was not until long after nationalization – in April
1966 – that British Rail was able to introduce a completely new concept
of train operation: following electrification from Euston to Liverpool
and Manchester, they began fast regular interval services with start-to-
stop average speeds as high as 71 mph (114 km/hr). The result was a
phenomenal doubling of traffic in only four years (ibid., pp. 15–17).
British Rail realized that they could effectively compete with air traffic
at timings of up to three hours between London and major cities. But
this would require very high-speed operation: much higher than the
Euston services.

As early as 1966, their engineers had identified two alternative ways
of achieving sustained speeds of 125 mph (200 km/hr). The first was to
use established traction and rolling stock, by reconstructing the track
and signalling systems on a few select lines with superior standards of
alignment. Unsurprisingly, the two obvious candidates proved to be
Brunel's line from London to Bristol and William Cubitt's line from
London to Doncaster – the latter, surveyed and built some ten years
later when locomotives had attained greater power, having been built
at great cost to a ruling gradient of 1 in 200 (ibid., pp. 19–20, 23;
Grinling 1898, pp. 28–9). The Brunel line had the additional advantage
that, thanks to late-19th-century duplication, it was four-tracked
continuously for 53 miles (85 km) from London to Didcot, allowing the
125s to be segregated from the slower-moving suburban traffic.

On lines not so perspicaciously engineered, BR engineers devised
the solution of a train that would tilt round curves: the Advanced

Passenger Train (APT). In the event, late in its trials it proved to have serious defects and will not now enter service – at least not in the near future. Meanwhile, Brunel's grand concept continues to demonstrate its effortless superiority a century and a half after it was begun.

The trains that run on it are, however, very different from those that Daniel Gooch designed at the new Swindon locomotive works. They are extremely light in weight – 383 tons, a saving of nearly 20% over the trains they replaced – and very powerful, with the capacity to accelerate from a standstill to 66 mph (106 km/hr) in 1 minute 39 seconds and to 125 mph (200 km/hr) in 6 minutes 49 seconds. They run on continuously welded rail and on a renewed formation and ballast, which Brunel – who experimented ceaselessly in these areas, but with notable lack of success – would only have envied. And they are controlled by new multi-aspect signals, although, because of their superior braking capacity, they required relatively little signal re-spacing (British Rail 1979, pp. 9, 18).

The critical geographical point is, however, that they run on Brunel's line. That line, by following the Thames, established the place of Reading on the railway map as surely as on that of the stagecoach era. By avoiding Newbury, it put that town into a rural backwater until the M4 – deviating from Brunel's route, and here crossing an improved A34 from Winchester to Oxford – helped return it to a key point on the late-20th-century pattern of communication. And finally, by establishing the new town of Swindon at the summit of the line, it fixed a further key urban node on the Western Corridor, the real significance of which would emerge over a century later. If the London–Bristol axis has a base in transport infrastructure, it deserves to be known not as the M4 Corridor, but as the Great Western Corridor.

Summing up:
the significance of infrastructure

The evidence in Chapter 5 shows that industrialists reckoned the Corridor's transport infrastructure as a significant factor in their locational choice. But, as emphasized there, a significant difference separates the small indigenous entrepreneurs from the big multinationals. The first group are there because they started there – M4 or no M4, Heathrow or no Heathrow. But, given that they are there, all three pieces of infrastructure are helpful and significant in varying degrees. The multinationals, in contrast, tended to arrive later and to base their locational choice fairly specifically on the Corridor's accessibility. For them, the infrastructure does appear to have been crucial.

That being so, the timing of these key public investment decisions is significant. The railway was long in existence, and Reading always provided a major first stopping point for many expresses – particularly after the development of a bus link to Heathrow in the early 1970s. All that the 125s did was to enhance this quality of service, investing Reading with an accessibility to London that was almost unique, in terms of times and frequency, among the towns of the Outer Metropolitan Area. This quality made the town's centre highly attractive for decentralized office development, and it was clearly a significant factor for administrative high tech headquarters offices. Similarly, Heathrow was in existence throughout the postwar period, and its unique role, as the world's leading international airport, without doubt proved the critical consideration for most of the multinationals.

The M4, in contrast, was not there – or rather, when many of the locational decisions were being made, very little of it was there. Of course, it was certain that it would be built soon, but, until the mid-1960s, it was not at all certain where it would be built. The many high tech firms that located in Bracknell, or in south Reading, or in Newbury, could have no guarantee that the highway would pass close to their doors; the contrary, in fact, given the ministry's preferences at the time.

The same, interestingly, goes in even greater measure for the later major addition to the region's infrastructure – the M25, London's orbital motorway. It was true that the motorway appeared as part of the national roads programme from the 1960s onwards, but without any firm date for construction. Even in the mid-1970s, when the first short stretches began to be built, its place was by no means secure. Only at the very end of the decade did government policy embrace it as the number one priority in the programme. Here, again, industrialists could not have made their decisions with any confidence as to the future pattern of accessibility.

In so far as there was a plan that integrated these elements, it was a singularly disjointed one. As seen in Chapter 10, between the 1944 Abercrombie Plan and the 1970 Strategic Plan there was no major plan for the region. Infrastructure planning proceeded as a set of separate exercises conducted by three different agencies: the Ministry (later Department) of Transport, the Ministry of Civil Aviation and its successor the British Airports Authority, and British Railways. All responded to what they saw as national – and in Heathrow international – traffic demands. There was no suggestion, anywhere, of a plan to develop a corridor of growth except for the South East Economic Planning Council's *Strategy* of 1967, which was never accepted by government – though it did provide one input into the 1970 *Strategic Plan's* proposal for a major growth area straddling the M3 and M4. In any case, many of the Corridor's companies were already in place by that time. So the Corridor, like Topsy, just growed, but grew as a result of heavy, if unco-ordinated public investment – by a process of disjointed incrementalism, without conscious understanding of its possible consequences.

PART IV *So what?*

12

Conclusions and speculations

It is time to conclude. First, we extract and highlight the key conclusions from the previous chapters. They answer the *what?*, *where?* and *why?* questions about events up to now; they show how the present shape of high-technology industry in Britain has evolved out of the events of the past half-century. Secondly, we go on more speculatively to suggest some implications, thus to try to answer the final question we have set ourselves: *so what?* Our main conclusions are eleven in number.

Some key conclusions: what? where? why?

What?

Our first conclusion addresses the question posed at the beginning of the book: what does the development of high technology industry actually mean in economic terms for Britain? We have established that – quite contrary to popular mythology – high tech in Britain has not *directly* been a generator of new manufacturing jobs, but a marginal contributor to general industrial decline. The seven industries defined here as constituting the high tech sector collectively lost 75 000 jobs between 1971 and 1981, and another 30 000 between 1981 and 1983: an overall loss of over 100 000, or 15% of the 1971 workforce. Within this list, only one industry – radio and radar capital goods – showed any job gain at all. A longer list of 43 high tech industries shed over 366 000 workers between 1971 and 1981. Significantly, a comparable list of industries in the United States gained well over 1 million workers in the shorter period 1972–81. This is a measure of the British failure to compete in the international high tech race. Nevertheless it is a race which Britain *must* dare to take part in because the stakes for the future are too high and the consequences of not competing are too dire.

But it has become increasingly evident to us as our research progressed, that it is important to locate high tech manufacturing within its wider *economic* context and, rather than treating it as an isolated activity in its own right, to begin to consider its relationship with the growing service sector. The links between the two is vitally important but poorly understood. The nature of this relationship will be considered in more detail later.

Where?

Our second conclusion answers the question, within the time period under review, where has high technology industry tended to locate, and can a geographical pattern be discerned? As well as considering high tech in its economic context, it is also important to locate high tech manufacturing within its *spatial* context. In so far as there has been growth in high tech employment, it has been concentrated in a few counties mainly in the outer South-East. Between 1975 and 1981, Berkshire gained 7600 new high tech jobs, Hertfordshire 5900, Hampshire nearly 2200, Surrey 1800 and Kent 1700. In the top eight counties ranked by absolute high tech job gains, there was only one non-South-East county, Clwyd. Of course, even in the South-East, the gains were modest – and they were massively counteracted by losses of 17 000 high tech jobs in Greater London and 4300 in Essex. Overall, even the South-East barely gained at all. But, as indicated earlier, within the fastest-growing South-East counties, the growth of high tech was totally overshadowed by expansion of the service industries, which are increasingly major users of high tech products.

Thirdly, both the pattern of high tech concentration and the pattern of high tech growth cast doubt on another popular myth: that of a lengthy, but isolated and discrete Western or M4 Corridor. Rather, the patterns suggest a Western Crescent of high tech which wraps around London from Hampshire in the extreme South-West, via Surrey and Berkshire, to Hertfordshire in the North-West. Within this crescent, individual centres or poles of high tech growth can be identified: Portsmouth, Bracknell, Reading, Hemel Hempstead. The eastern end of the M4 – a short M4 Corridor – in particular, is revealed as an area of intense economic activity superimposed on top of the Western Crescent. The typical high tech firm within the Crescent, and especially in the eastern M4 area, will be relatively new, relatively small, and relatively biased towards administration and research, rather than pure manufacturing functions.

Why?

The remaining conclusions address the fundamental question, why? Why has this type of development occurred at this time, in this place, at all? In order to answer this question it is first necessary to summarize our findings, based on reviewing one by one the three original explanatory hypotheses put forward in Chapter 6.

The initial hypothesis, which was termed locational displacement, suggested that during the course of the profit cycle older private sector firms would spawn newer, smaller offspring which would locate near their parent organizations, but which would gradually move away from the major existing urban centres.

Our fourth conclusion, drawing on interviews with firms and some historical study, suggests that the high tech concentration in the Western Crescent represents a continuing trend. This is a spiralling outward movement of a much older high tech sector, which was located in north-west London in the 1920s and 1930s, and which in turn represented an outward movement of small firms originally based in the heart of London, in places like Clerkenwell, Pimlico and Lambeth. This is consistent with London's overall continuing economic and population decentralization. But the process did not simply or mainly consist in the outward movement of individual firms in search of space, compelled by postwar planning regulations to leapfrog London's green belt; it also consisted in a process of new firm formation, often through spin-off from existing larger firms. Here, very large established firms, generally London-based, played a critical role. Thus, some form of local displacement does seem to have taken place, but with a strong spawning of new enterprises as well as decentralization of established ones.

The next explanatory hypothesis put forward in Chapter 6 suggested that the real reason for high tech growth in the area was related to public expenditure on research. The hypothesis suggested that organized research, closely associated with government-financed R&D, particularly following World War II, has played a major role in attracting large, well-established high tech firms and in helping to generate small, new, innovative firms as well.

The fifth conclusion, therefore, suggests that the process of local displacement from London alone cannot satisfactorily explain why the resulting spatial pattern came to have such a pronounced western and southwestern bias. A critical factor here was the presence of the public sector, in particular the defence research establishments, in these locations. DREs were already there before World War II, though they massively expanded both during the war and again during the cold war period of the 1950s. The reasons for this geography go back into history, and lie in the traditional concentration of the defence establishment in centres like Portsmouth from the 16th century and Aldershot from the 19th – in areas accessible to, but protected from, the traditional wartime enemy, continental Europe. Purely contingent British factors, like the presence of old royal mansions in the area west of London, also played a role.

The organized research hypothesis is also supported by the next finding. Our sixth conclusion is that the DREs played such a crucial role because of the rapid buildup of military electronics in the 1950s and 1960s. Development contracts were placed with a limited number of key suppliers. Links with them, on a day-to-day and face-to-face basis, were intense. They developed prototypes, and then invariably obtained the subsequent production contracts. Subcontractors clustered around them, including firms founded through spin-offs from the parent firms. All this created intense agglomeration economies, leading to the creation of a new, wider-spread regional version of the traditional inner city industrial quarter. Thus a disproportionate part of military procurement expenditure came to be concentrated in southeast England and neighbouring parts of the south-west region, ironically counteracting and swamping the effects of conventional regional development expenditure. Massive defence expenditure on aerospace and electronics continues to sustain a considerable proportion of high tech activity in the Western Crescent. The continued importance of face-to-face contacts is likely to ensure that the geographical concentration persists.

The seventh conclusion relates to the last of our three hypotheses and concerns the role of multinational corporations. The 'international capital' working hypothesis suggests that there is an increasing tendency for international capital generally, and international high tech capital in particular, to organize its global activities spatially in such a way as to maximize returns consistent with its production requirements. There is little doubt that overseas high tech companies have played a major role in the development of the Western Crescent and especially the eastern end of the M4 Corridor. Most of the large companies now represented in this area are overseas and particularly American companies. The presence of these overseas multinational companies only explains, however, the continued growth of high tech in the area, rather than its genesis. Such companies have featured largely only since the 1970s, by which time the area was already established, albeit not popularly recognized, as a concentration of high tech activity.

In this sense, then, our findings do support the international capital hypothesis, but with the proviso that rather than being responsible for the development of the high tech concentration, these multinational corporations have taken advantage of a location that had already been established indigenously. The location has in turn been used, of course, not as a peripheral, low-cost production centre, but as a continental or national management/R&D/marketing annex to corporate headquarters. The area fits the hypothesis still further because many of the larger companies carrying out these functions in the M4 area do, in turn, carry out production in more peripheral, low-cost areas in the UK (e.g. Digital in Scotland) or other parts of the world. Thus, the general features of the standard hypothesis do seem to hold.

There are, however, important exceptions to the general hypothesis. As mentioned earlier, Racal, one of the few UK-based electronics multi-nationals, has a very different corporate spatial structure, which focuses all corporate activities in relatively small autonomous units in single locations. Many of these units are in the eastern M4 Corridor area.

Multinational companies have, then, made a major contribution to the growth of high tech activity in the M4 area, but can be held responsible for sustaining rather than starting the phenomenon. They have, however, had a major role in developing the area as a 'core' location in the national and international geography of high tech – with relatively little large-scale production, but considerable management, research and marketing functions. This status is likely to have significant long-term implications for the area; it might provide a more robust industrial base than is available in other currently buoyant but structurally fragile production locations.

Our eighth conclusion, following this review of the three original hypotheses, is that the growth of the high tech firms in the Western Crescent cannot be accounted for by any *single* factor, event or explanation. Life, as usual, is more complex than that, particularly in the case of high tech industry which is, by its very nature, both diverse and dynamic. As a result of conducting our research, we know that each of the hypotheses examined above (locational displacement, organized research and international capital) are correct at different points in time, at different places within the Western Crescent.

As Chapter 7 demonstrated, a substantial number of historic high tech firms have decentralized from London and a number of smaller firms have also been established in the area. At the same time the growth of the government research establishments, detailed in Chapter 8, proved to be a major source of attraction for a variety of reasons to private sector firms both large and small. Finally, it is clear that a growing number of multinational firms are now established in the Crescent which have been attracted by the same factors which initially caused both the indigenous firms and the GREs to locate there. Thus, not only are the explanatory hypotheses simultaneously although spatially selectively true, but they are mutually and positively reinforcing.

Despite their reinforcing nature, however, we should accept that while these different theoretical perspectives help to explain the development of the phenomenon over time, it is unnecessary, and possibly unwise, to attempt to construct a single overarching and eclectic explanation, as suggested earlier in the book, purely for the benefit of superficial neatness. In attempting to do so, we would run the risk of sacrificing a necessarily rich and diverse, if slightly chaotic, explanation for the sake of imposing conceptual order of a very procrustean kind.

We are confirmed in this view by the discovery of an important, but largely overlooked, element of explanation which did not seriously figure in our original working hypotheses, or in most of the theoretical literature from which these hypotheses were drawn: the vital role, in the broadest sense, of the public sector.

Our ninth conclusion is that the growth of the Crescent and eastern M4 Corridor has been powerfully aided by important public sector policies in addition to defence expenditure. Perversely, this aid has occurred by accident. National transport planning priorities after World War II caused Heathrow Airport to be established west of the capital to serve as the nation's main international traffic centre; in the 1960s, the M4 London–South Wales motorway was routed south of Reading; in the 1970s, British Rail introduced the first high-speed Inter-City trains on this route. None of these policies was supposed to add to the development of industry in this corridor; all did. A consistent response in our discussions with high tech managers in Berkshire was the crucial importance of the area's excellent communications infrastructure in their decision to locate and remain in or near the county.

A tenth and closely related conclusion is that land use planning – particularly strategic planning – has played a significant but curiously ambivalent role in the development process. At one level it has actively fostered growth in specific locations through the designation of new towns, such as Bracknell in 1949. And it has also singled out major growth zones, including the Reading–Wokingham–Aldershot–Basingstoke complex, for example (Area 8), which was proposed in the *Strategic plan for the South East* in 1970.

At the same time, and paradoxically, it has also done a great deal to limit urban expansion – particularly of Greater London – and to preserve both areas of outstanding natural beauty and land of high agricultural value. It has acted to discourage new development generally, powerfully assisted after 1955 by government green belt policy. The cumulative result of this effort over a period of decades is a limited number of linked, medium-sized growth areas set in a sea of general rural restraint. As a result of what could be termed benign perversity, it is just this policy of protecting the character of the countryside which made the area so environmentally attractive to both indigenous and incoming high tech firms, especially American multinationals.

Our eleventh and final conclusion, then, is that it is crucially important to make the point that the common thread which runs throughout these conclusions is the diverse, often unrecognized, but extremely significant role played by the public sector – both civil and military – over a period of decades in providing the type of pre-conditions so necessary for the development and sustained growth of high tech industry. But once again there are a number of paradoxes inherent in this process. The fact that so many private firms, with such a large degree of locational choice, have deliberately *chosen* to site themselves within the Western Crescent and particularly the eastern end of the M4 Corridor is no accident. But if public policy is supposedly based on co-ordinated conscious decisions coupled with a lively awareness of the *consequences* of the decisions made, the fact that the area has developed in the way that it has, with the speed that it has, is almost entirely accidental.

The public sector has played a crucial role in the high tech development of the Western Crescent, but in a fragmented, narrow, unconscious, and even on occasion contradictory way. It is certainly the case that major public investments were made in research establishments, major airports, new motorways, and rapid rail infrastructure, but these decisions were taken by different ministries on departmental grounds with an extremely limited awareness of their cumulative spatial consequence. Ironically, the very activity which could have performed this role, strategic planning, was by the 1980s in precipitate decline, possibly at the time when it was most required.

There is little doubt that without the types of publicly provided fixtures and facilities described above, it is extremely unlikely that the area would have proved as attractive to advanced private sector firms as it has. The public sector, over the past several decades, has invested billions of pounds and created tens of thousands of jobs directly within the Crescent through capital construction programmes, and the employment and training of workers. For a number of reasons, discussed earlier, the private sector has found the physical transport infrastructure of great value, but perhaps as important, although certainly less discussed, is the *knowledge infrastructure* which has been funded by the public sector, particularly in the GREs.

This publicly funded knowledge infrastructure built up over time has allowed the area to develop and grow, and private firms have substantially benefited from it in terms of innovative products and inventive people, in the form of a skilled and highly trained workforce. The suggestion which is sometimes advanced in Britain that high tech development is a shameless triumph of the private sector unaided, or that the future of high tech development can be left exclusively to market forces, is a dangerous view which is not shared by the governments of the United States, Japan, West Germany or France in policy or practice.

The fact that the Western Crescent *has* developed, although in an admittedly unco-ordinated way, could be interpreted as an argument for further reductions in investment by the public sector, as attempts are made to stimulate, or even replicate, high tech growth in other parts of Britain. The fact that the M4 Corridor developed largely by chance, it could be argued from this point of view, means that it could happen again in other areas by the same mechanism.

This argument is extremely dubious for two main reasons. Firstly, however unco-ordinated the individual public sector investments in the Western Crescent were, they nevertheless provided a crucial financial, physical and technological foundation for the area's development. Secondly, Britain's major trading competitors in the field of high tech are deliberately and consciously fostering advanced industrial development as acts of national policy, and they are backing this policy with the commitment of substantial resources. This point will be returned to later as speculations are advanced about the future of high tech development in Britain.

So what? Some implications

As a result of reviewing these findings in total, based on a substantial body of evidence and analysis about the recent development of high technology in Britain, the ultimate test of the value of the research remains: so what?

Our aim must be to judge the significance of the recent emergence of the Western Crescent, and particularly the M4 part of it, as a centre of high tech activity. First, however, because we have placed this area within a much broader context, we might dwell for a moment on the significance of UK high tech activity. To cut a long story short, at the national level the picture is a source for concern. The popular assumption that the high tech manufacturing sector has been on its own a major generator of jobs is a myth. Our findings are consistent with more informed judgements, which stress the relatively poor UK performance in high tech manufacturing. It has produced directly relatively few jobs; a tiny fraction of those that would be needed to compensate for the massive losses in older smokestack industries. One point on which popular and informed opinion agree, however, is that these high technology industries are crucially important. In principle, we might think of alternative futures (Massey 1985), but in practice we are bound up in an international economy that places increasing weight on both the production and the use of high tech products. The general view is that a strong representation in such industries is essential for future economic success.

If this is true for nations, it is likely also to be true for localities. As our review of the geography of high tech activity in the UK has shown, there is some high tech growth in a small number of areas and considerable decline in many others. The net effect is our depressing national picture.

In many ways the finding about the geography of high tech in Britain is currently more worrying than our conclusions about national performance. This is because all the evidence suggests that the familiar regional pattern of economic advantage and disadvantage in Britain is now being reinforced, not changed, by high technology activity. The South-East, South-West and East Anglia have gained a double benefit: they have taken the large majority of the few new high tech jobs, and these jobs tend to be in the more durable non-manufacturing activities.

How, then, do we judge one of the few localities – the M4 Corridor superimposed on the Western Crescent – that has stood out against this gloomy backcloth? Just how bright has been the 'Western Sunrise'? The answer is, it depends. It depends on the performance criteria adopted, and whose interests we are considering.

Job growth

Clearly, a major criterion is jobs. Judged from this perspective, we might conclude that even those areas with the highest growth in high tech employment have had only modest gains. Berkshire, for example, at the heart of the M4 Corridor, had a growth of only a few thousand high tech jobs in the decade up to 1981, and many of these were in overseas and particularly American companies. Also, the jobs have tended to be in skilled occupations, providing few opportunities for unskilled and semi-skilled workers made redundant in the traditional manufacturing sectors. Thus, even in the most fortunate county, the *direct* contribution of high tech to employment, welcome as it is, has been modest. The service sector has provided many more jobs over the same period; in a ratio of about ten to one with high tech production jobs.

Should we conclude, then, that the job-creation possibilities of the high tech sectors are grossly overstated? In direct terms, the answer must be yes; at least in a British context. In more general, direct and indirect terms, we cannot be so definite. It may well be that some, and possibly many, of the service sector jobs and even some non-high tech manufacturing jobs in Berkshire result indirectly from its high tech manufacturing. But what evidence is there to suggest such a link between high tech and service jobs?

Not so long ago, economic geographers and regional economists were prone to treat all growth in service employment as derivative from the demands generated by goods-producing industries. They no longer make that mistake: for the City of London, and indeed for the British economy as a whole, the export of financial services is a basic industry just as much as steel was once for South Wales, or cars for Coventry. Part of the growth of office-based services, in Crescent towns like Reading and Basingstoke and Slough, appears to represent a local deconcentration of just that kind of activity.

But a doubt remains. Jonathan Gershuny has argued that if a proper input–output table were drawn up, the growth of output and employment in the so-called intermediate producer services is derived from manufacturing, primary or other service industry. The reason, he suggests, is that as physical productivity in the actual manufacturing process has increased, an ever-increasing proportion of the total activity of the manufacturing sector has taken the form not of production but of specialized skills – many of which are contracted out to other agents (salesrooms, shippers, advertising agents, accountants) who are classified as providers of services:

> The growth of intermediate producer service-industries has been
> a consequence of the search for efficient production through the
> division of labour. Instead of employing particular sorts of
> specialised service labour, firms sub-contract work to specialist
> service agencies. By doing so, they can gain advantages of scale
> and specialisation and frequently, of close supervision of work
> that would not otherwise be available. This may be expected both
> for professional and technical services (e.g. design, advertising,
> marketing, technical consultancy) and for other less highly-skilled
> services (e.g. cleaning, catering, security). (Gershuny & Miles
> 1983, p. 49)

Viewed in this light, at least part of the growth of business services in
the Crescent might reflect demand from the high tech manufacturing
base. There is no way of saying, without undertaking a complex local
input–output study which is entirely beyond our research design or
our available resources. We can, however, speculate from some
indirect evidence. Researchers at the Department of Employment have
tried to draw up an input–output table for the UK economy, including
services, for the year 1972 (Robertson *et al.* 1982). It is necessarily quite
crude; it aggregates producer services, and a number of personal
consumer services, into a very broad category of 'miscellaneous
services'.

It suggests that about one-fifth (to be precise, 20.7%) of employment
in these services depended on demand from manufacturing, as against
36.6% depending on consumers' final demand, 12.7% on public
administration, and 11.1% on exports (ibid., Table 77). Put the other
way round, a £1 million increase in final demand for manufacturing
industry is calculated to generate 327 extra jobs, only 27 of which
would be in the miscellaneous services category. In contrast, £1 million
injected into final demand for miscellaneous services would generate
341 extra jobs, of which 40 would be in manufacturing and 279 in
miscellaneous services (ibid., Table 93). These results, for all their
admitted limitations, suggest strongly that at the national level the
miscellaneous service sector is not very dependent on manufacturing
industry.

Locally, of course, the relationships could be different. Rapid growth
of new manufacturing industries might call forth a demand for
specialized service functions that needed to be performed locally. But
this seems on the face of it unlikely. Such functions would normally
continue to be highly concentrated in very high-level service centres,
especially London. The only exception would occur when local
demand reached some critical mass, justifying the development of a
detached sub-centre of activity. Some anecdotal evidence suggests that
this may recently have happened with specialized accountancy and
management consultancy services in the Reading area. But it is
doubtful whether this could be quantitatively very important.

There is, however, another more likely possibility: that high tech manufacturing growth brings with it a demand for software development and specialized servicing functions. Indeed, the core of Gershuny's argument is that software, in the widest sense, will provide the source of most new jobs through social innovations in the way that consumers obtain their final service functions (Gershuny & Miles 1983, pp. 237–40). It seems possible that the employment multiplier effects through software development could be particularly large in those areas of manufacturing high tech where the product itself is highly innovative and experimental, as seems to be the case for the output of many firms in the Crescent. So even if few direct manufacturing jobs are created, there might be a considerable growth through the associated labour-intensive service sector. This, though it cannot presently be proved, is the hope.

Despite these views, we are unable to say anything definite at present about the indirect employment effects of high tech manufacturing. We can observe that areas such as Berkshire have had a coincidental growth of high tech on the one hand and producer and consumer services on the other. If there is cause and effect between the two, we are not entirely sure of the direction of causation, never mind the magnitude. Most reviews of this issue tend to take a clinical, economic perspective. There may equally be an important psychological element involved. There is little doubt that Berkshire's high tech activity, as small as it is in direct job terms, has given the area a certain cachet that has in turn attracted investors across a whole range of activities; the attitude and influence of property agents is a case in point. The same high tech 'culture' may now be observed in, for example, Swindon and Cambridge.

Job structure

Another criterion we might use in judging the Crescent and Corridor is the contribution that high tech makes to the provision of jobs across the skill range. These areas do, in fact, offer an extreme illustration of the mismatch of labour supply and demand. Company after company in our interviews in Berkshire reported that their chief problem was the shortage of properly qualified labour and the difficulty of recruiting it, exacerbated by the high costs of housing compared with the rest of the country. Indeed, a number were suffering from extraordinary numbers of vacancies. Yet these interviews took place at a time when local unemployment rates, though low by national standards, had reached historic high levels. In May 1984 the rate for Bracknell was 6.7%, for Basingstoke 6.9%, for Slough 10.0% and for Reading 10.6% (SERPLAN 1985, p. 40). Employers were looking for the kinds of specialized skills that the unemployed had not got. And the paradox is that, with unemployment at a post-World War II high, the recovery of the local economy was being held back by labour shortages.

The evolution towards a software-based service economy, on the Gershuny model, could only deepen this paradox. Such an economy would place an even greater premium on special technical skills. Though its growth might also generate – through income multipliers – a demand for low tech jobs in consumer services like shops and restaurants and personal services, this would depend on the skills at the base of the pyramid. On the evidence so far, there is no guarantee that the educational system is providing this, either locally or nationally.

Again, then, we come back to the indirect effects of high tech activity. It may provide low skill jobs in the service sector, but only if it generates a service sector.

Job stability

A further criterion we might consider in asking the question *so what?* about the Crescent and Corridor concerns the durability of the high tech phenomenon. We have provided evidence of three reasons for assuming that the growth has deep and stable roots. The first is the longevity of the phenomenon; it has deeper historical origins than is popularly assumed. The second is the particular role that the area plays in corporate spatial structures; it acts primarily as a 'control' administrative and R&D location. Such functions are likely to be more durable than production. The third reason is the close association of many of the high tech companies with the local DREs. Despite ministerial pronouncements about decentralization of research establishments, this locational association is likely to persist.

In considering the future durability of high technology industry in the Crescent, two doubts come to mind. One is that the defence connection *may* not be as lucrative in the future as in the past. Defence expenditure in the UK has risen sharply in real terms in recent years, but is now budgeted to decline. The other doubt is that an increasing share of high tech jobs, certainly in Berkshire, are in American companies. Such jobs are often regarded as more vulnerable than those in indigenous companies. In fact, the evidence to date in the area suggests that the opposite is more likely to be the case.

Weighing up both the optimistic and pessimistic evidence about durability, then, gives us good reason to assume that the growth of high tech in the Crescent, and certainly the Corridor, is not a transitory phenomenon. The advantage the area has gained will continue for some time.

Judged critically from the perspective of these criteria – job growth, job structure and job stability – we can find a few clouds obscuring the Western Sunrise. But we must be fair and realistic. Judged from a perspective that includes the UK generally, the Western Sunrise is clear and bright. The area has an economic vitality virtually absent elsewhere in the country; new jobs have been provided, and continue to be provided; new high tech firms continue to move into or start up in the area; these firms are managing to compete in the crucial high tech markets. Particularly successful parts of the Crescent, such as Berkshire, are looked upon with envy by other local authorities. They would dearly love to emulate its success.

Replicating the M4 success

Although the direct employment growth attributable to high tech in the M4 Corridor is very modest, imagine what the national picture would be if all 64 counties in Britain had had the same absolute gain. If we multiply up Berkshire's 8000 job growth from 1971 to 1981, the equivalent national growth would have been 512 000 new jobs – virtually equivalent to the level achieved in the United States. This is a rather nonsensical calculation, of course, but it serves to make the point: until we have localities doing well in the high tech stakes, our national performance in this sector cannot be other than dismal. At the end of the day, national performance is the sum of local performance.

This is why it is vitally important that we understand why certain areas, and the eastern M4 Corridor exemplar in particular, have had some success in attracting and nurturing high technology companies. The prospects for replicating this success in other locations has been one motive for the attempt here to provide some explanation. If our analysis is correct and the public sector, however unwittingly, has had a major role in the M4 Corridor phenomenon, then we have more reason to assume that replication could be attempted.

But where? Which areas are most likely to respond to some attempt at replication of the M4 success? It would be nice to think that some of our depressed regions might benefit from such endeavour. Indeed, they might; some of the lessons learned from our attempted explanation of the M4 Corridor growth could at least increase the prospects of some currently depressed regions. There is little doubt, however, that if these lessons are to be used to improve our national performance in the high tech sector, the best bet would be to concentrate on areas that already have some of the required features. This is likely to mean extending the Western Crescent.

In so far as public investment in national transport infrastructure has aided the development of the Crescent and Corridor, how far could it do so again? The irony is that Abercrombie's great cartoon-like vision of 1944, for the development of the Greater London region, will be realized only in late 1986 with the completion of the M25, the 117-mile (188 km) long orbital motorway encircling the London conurbation and tying together the settlements in the outer parts of the south-east region.

Expert analyses (Lichfield 1981, SERPLAN 1982) agree in predicting that the completion of the motorway will have major impacts on pressures for development, especially in the favoured western sector from the A3 Portsmouth radial to the M40 Oxford radial, and especially just outside the M25 at the outer edge of the green belt. The Standing Conference planners recognize this pressure, but hope to channel it into the central Berkshire – north-east Hampshire growth zone and into the Milton Keynes area, which are both identified in structure plans as areas for substantial new development. But they also hope that the enhanced orbital accessibility, which completion of the M25 will bring, will prove particularly helpful to the relatively deprived eastern sector between the A1 Stevenage–Peterborough radial and the A127 Southend radial and, even more, between the A127 and the A2 Dover radial (ibid., pp. 52–3). The development of Stansted airport north-east of London, and of the Channel Tunnel to the south-east, could further assist this development, perhaps creating new radial corridors of development in areas which are now in need of it.

Here there is strong supporting evidence from the Transport and Road Research Laboratory, whose research suggests that dramatic gains in accessibility will occur as a result of the M25 along the M11 Corridor from London to Cambridge – greater, indeed, than in any other radial corridor leading outwards from London. Employers along this corridor will gain over 100 000 potential workers within a 30-minute drive time; around Harlow itself, the gain is more than 200 000 workers (Jones 1982, Figs 17, 18). All this strongly suggests that, just as government investment in the 1960s aided the development of an M4 Corridor, so in the 1980s and 1990s it could contribute to the formation of an M11 Corridor (Breheny *et al.* 1986). And the government's decision in 1985 to develop Stansted as London's third airport half-way along this corridor could prove analogous to the crucial Heathrow decision of 1945.

The main reason why this might be true is that there is already some evidence that the area around Cambridge has begun to demonstrate some of the same features that earlier characterized successful American high tech areas like Silicon Valley or Boston's Highway 128. The study by the consultants Segal Quince and Partners, *The Cambridge phenomenon*, identified some 350 such companies, most of them small, that had developed in and around the city since the late 1960s – 100 of them since 1980 alone (Segal Quince 1985). And, quite unlike our findings for the Western Crescent, most were founded by people who were at the university or who worked for other small companies in the area. It concludes that a critical factor is the existence of a large number of loose social networks that allow the technology-based companies to keep in touch with university research, and with each other – exactly the feature that characterized the growth of Silicon Valley. The absence of large established multinational companies has helped, for such firms are less likely to breed entrepreneurs. And Cambridge has its own equivalent of the Stanford Industrial Park, out of which Silicon Valley grew: the Cambridge Science Park to the north of the city, developed by Trinity College, which now houses more than 40 companies.

Many of these firms, including those best known to the public (Sinclair, Acorn) are in information technology. But others, so far less known (Cambridge Life Sciences, IQ Bio, Cantab) are engaged in biotechnology research. For this, the area has an outstandingly strong foundation. Thus the real role of the M11 Corridor might well be in biotechnology, an area now roughly in the same state of development as information technology was in the 1950s.

This is particularly important, because of the hope that the M11 Corridor could emulate the success of the M4 Corridor. It could become a genuinely entrepreneurial subregion, developing new technologies for commercial purposes and acting as the basis for a new industrial revolution in which Britain, by reason of its scientific excellence, could play a leading role.

The missing link:
a new type of strategic planning

If replication of the eastern M4 phenomenon is desirable, and areas like the M11 Corridor already have certain advantageous features, what needs to be done to ensure that their promise is fulfilled? Now that we know a good deal more about the preconditions for high tech growth, and indeed about different forms of high tech, including electronics, aerospace and biotechnology, how should this knowledge be applied in our attempt at replication? Naturally, we look to the public sector both to apply such knowledge and to intervene as necessary; apparently all the more so in this case because of the crucial role, explained here, that the public sector has played in the development of the M4 concentration of high tech.

The technological growth in the Western Crescent, largely since World War II, has been partly the result of historical decisions made by separate sectoral public agencies, without any benefit of co-ordination, and above all without any recognition of their impact on regional or urban development. The defence agencies expanded their research establishments, the establishments created a network of contractors and subcontractors, the planners forced expanding contractors out of London and over the green belt, the transport planners built airports and roads and high-speed railways. All unwittingly contributed to the growth of Corridor and Crescent; none consciously worked to this end, and some – the local land use planners – believed that they were working to stop it. Yet paradoxically even they helped, all the same.

Although the consequences of these various public sector decisions were not foreseen, they were at least made at a time, in the 1960s in particular, when strategic planning was politically acceptable and was actually practised. Now, the whole concept of this type of conventional strategic planning has been officially condemned as outmoded (GB

Department of the Environment 1983), and the Berkshire County Planning Officer reports that in the view of his councillors strategic planning is 'unnecessary, ineffective, bureaucratic, and unpopular' (Stoddart 1983, pp. 150–1). In short, strategic planning in the UK has virtually disappeared, both in principle and practice.

This condemnation, it can be argued, is based on a misconception that sees the role of strategic planning only in terms of rigid blueprints to produce short-term results, whereas its chief role should be to provide a longer-term framework to guide the processes of change, by bringing the public and private sectors together for future investment decisions.

Associated with this is another popular misconception: that the value of strategic planning lies only in its positive proposals for new development. In fact, almost as valuable is the negative environmental protection aspect: good long-term strategic planning can also prevent undesirable developments, thus protecting the environmental qualities of an area. And this is what has been achieved, willy-nilly, in the case of the core of the Corridor. This success can be contrasted with the saga of Silicon Valley, where precisely the lack of such control has led to a steady deterioration of the area's environmental quality, making it less attractive as a high-tech centre (Saxenian 1985a, b).

Notwithstanding these misconceptions, the fact is that strategic planning is profoundly unfashionable in the UK at present. Thus, if our logic is reasonable – high tech development is crucial, the M4 success is worth replicating, there are areas in which such replication might work, and the public sector would have to play a major role – we have a problem. In fact, we have a double problem. Not only do we have the obvious difficulty of using our knowledge of the M4 success in developing other areas, but we have a problem of means: public sector intervention seems to be crucial, but unavailable.

All of this seems paradoxical because the fostering of high tech growth is the declared policy of the present government. It has stated, 'High technology industries are essential to the country's future prosperity', and has urged local planning authorities to cater properly for their needs (GB Department of the Environment, 1984). The government does not recognize the paradox, however, because presumably it assumes that the private sector alone has created whatever high tech success we have had and can continue to do so. It is hoped that we have demonstrated that this assumption is incorrect and is not supported by the facts.

Without some positive strategic planning we are likely to see only a slow, accidental and randomly located (except that we can be sure it will be in southern England) growth of high tech in the UK. This will be reflected in the continuation of the dismal national performance relative to our international competitors. The alternative is to develop new forms of strategic planning for the next decade which increase co-operation between public and private sectors – particularly with regard to co-ordinating major investment decisions and their spatial impact; as well as deliberately fostering what we have earlier termed the knowledge, as well as the physical, infrastructure so that these new areas possess the preconditions for growth.

Suitable sites in selected regions need to be identified: sophisticated training and re-training programmes introduced; and government-supported pure and applied research programmes increased at GREs and universities. As Britain's major trading competitors, who are in the process of doing much of this, are aware, such development cannot be done on the cheap. But it can be done if the will is there. Otherwise it could be left to chance once more.

At one level, then, high technology development is very much a national issue. The crucial importance of high technology to Britain's future is reflected at this level in the number of bodies, official and unofficial, who are considering and passing judgement on the country's prospects. Thus, we have government departments, House of Commons and House of Lords committees, special committees of inquiry, university research teams, and many more all considering the issue. We have more direct initiatives charged with the responsibility for either investigating high tech performance or with promoting new ideas and priorities. The common feature of all these efforts is that they are almost totally aspatial; in effect, they assume that all of this activity takes place on the head of the proverbial pin.

At the subregional and local level, however, as a result of this and related studies, we now know that geography matters. Spatial factors can, of course, only account for a fraction of our national high tech performance; but marginal influences are important in such a crucial sector, and, as we have stressed, national performance can only be the sum of local performances. If we wish to replicate the success of the development of high tech in other areas, these largely local lessons need to be studied by central government decision makers and national policy altered in appropriate ways.

We must learn, then, from those localities – and specifically from the Western Crescent, and its core eastern M4 Corridor in particular – where performance has been good. Possibly, if the will and the means can be found, other areas will then feel the spreading warmth of the Western Sunrise.

References

Abercrombie, P. 1945. *Greater London plan 1944*. London: HMSO.

Allen, G. C. 1929. *The industrial development of Birmingham and the Black Country 1860–1927*. London: Allen & Unwin.

Allen, R. 1983. *Major airports of the world*. London: Ian Allan.

Allward, M. 1966. *London's airports: Heathrow: Gatwick*. London: Ian Allan.

Altshuler, A. *et al*. 1984. *The future of the automobile: the report of MIT's International Automobile Program*. London: Allen & Unwin.

Angus, R. 1979. *The organisation of defence procurement and production in the UK*. Aberdeen Studies in Defence Economics, 13. Aberdeen: University of Aberdeen.

Ball, N. and M. Leitenberg (eds) 1983. *The structure of the defence industry*. London: Croom Helm.

Barnaby, F. 1982. Microelectronics in war. In *Microelectronics and society*, a Report of the Club of Rome, G. Friedrichs and A. Schaff (eds), 243–72. Oxford: Pergamon Press.

Bates, B. A. 1954. *Some aspects of the recent industrial development of West London*. Unpublished M.Sc. thesis, University of London.

Bell, D. 1974. *The coming of post-industrial society*. New York: Basic Books.

Bell, D. 1979. The social framework of the information society. In *The computer age: a twenty year view*, M. L. Dertouzos (ed.), 163–212. Cambridge, Mass.: MIT Press.

Berkshire, Administrative County of 1951. *Development plan: report and analysis of survey and plan*. Reading: Administrative County of Berkshire.

Berkshire, Royal County of 1978. *Central Berkshire structure plan: submitted to the Secretary of State for the Environment, June 1978*. Reading: Royal County of Berkshire.

Berkshire, Royal County of 1982. *Survey of employers in Berkshire 1981*. Reading: Royal County of Berkshire.

Berkshire, Royal County of 1985. *Survey of employers*. Reading: Berkshire County Council.

Berry, B. J. L. 1970. The geography of the United States in the year 2000. *Institute of British Geographers, Transactions* **51**, 21–53.

Birch, D. L. 1979. *The job generation process*. MIT Program on Neighborhood and Regional Change. Cambridge, Mass.: Massachusetts Institute of Technology.

Blackaby, F. 1983. The military sector and the economy. In Ball and Leitenberg 1983, 6–20.

Blue Book, The 1938. *The blue book: the electrical trades directory and handbook 1938*. London: Benn.

Bluestone, B. and B. Harrison 1982. *The deindustrialization of America: plant closings, community abandonment, and the dismantling of basic industry*. New York: Basic Books.

Boddy, M. and J. Lovering 1984. High technology industry in the Bristol city-region. Paper presented at the Sixteenth Annual Conference of the British Section of the Regional Science Association Annual Conference, University of Kent, 5–7 September 1984.

Bolton, R. 1980. Impacts of defense spending on urban areas. In *The urban impact of federal policies*, N. J. Glickman (ed.), 151–74. Baltimore: Johns Hopkins University Press.

Boretsky, M. 1982. *The threat to U.S. high technology industries: economic and national security implications*, mimeo. Washington, D.C.: International Trade Administration, US Department of Commerce.

Boyd, D. 1975. *Royal Engineers*. British Regiments Series. London: Leo Cooper.

Brady, T. and S. Liff 1983. *Monitoring new technology and employment*. London: Manpower Services Commission.

Braun, E. and P. Senker 1982. *New technology and employment*. London: Manpower Services Commission.

Breese, G. W., R. J. Klingenmeier, H. P. Cahill, J. E. Whelan, A. E. Church and D. E. Whiteman 1965. *The impact of large installations on nearby areas: accelerated urban growth*. Beverly Hills: Sage.

Breheny, M. J. and R. W. McQuaid, 1984. *The genesis of high technology industry in the M4 Corridor: a preliminary view*. Paper presented at the sixteenth annual conference of the British Section of the Regional Science Association, University of Kent, 5–7 September 1984.

Breheny, M. and R. McQuaid 1985. *The M4 Corridor Patterns and Causes of Growth in High Technology Industries*, Geographical Paper 87. Reading: Department of Geography, University of Reading.

Breheny, M., P. Cheshire and R. J. Langridge 1983. The anatomy of job creation? Industrial change in Britain's M4 Corridor. *Built Environment* **9**, 61–71.

Breheny, M., D. Hart and P. Hall 1986. *Eastern Promise? – Development Prospects for the M11 Corridor*. London: Derrick Wade and Waters.

British Rail 1979. *Rail 125 in action*. Bristol: Avon-Anglia.

Brook, C. A. 1982. High technology industry and science parks: What are the policy and development control implications? *Journal of the Royal Town Planning Institute* **68**, 180–2.

Brook, C. A. 1983. *The rapidly changing field of high technology development and 'science parks'*. Paper given at the Land Policy Conference, Oxford Polytechnic, March 1983.

Buswell, R. J. and E. W. Lewis 1970. The geographical distribution of industrial research activity in the United Kingdom. *Regional Studies* **4**, 297–306.

Butler, S. M. 1981. *Enterprise zones: greenlining the inner cities*. New York: Universe.

Canada Department of Defence Production (DPP) 1964. *Production sharing handbook*, 3rd edn. Ottawa: DPP.

Carus-Wilson, E. M. 1941. An industrial revolution of the thirteenth century. *Economic History Review* **11**, 39–60.

Castells, M. (ed.) 1985. *High technology, space, and society*. Urban Affairs Annual Reviews, 28. Beverly Hills and London: Sage.

Checkland, S. 1975. *The upas tree*. Glasgow: Glasgow University Press.

Chew, V. K. 1981. *Talking machines*. London: HMSO.

Chinitz, B. 1960. Contrasts in agglomeration: New York and Pittsburgh.

American Economic Review, Papers and Proceedings **51**, 279–89.

Christaller, W. 1966 (1933). *Central places in southern Germany*. Translated by C. W. Baskin. Englewood Cliffs, N.J.: Prentice-Hall.

Clapham, J. H. 1910. The transference of the worsted industry from Norfolk to the West Riding. *Economic Journal* **20**, 195–210.

Clark, J., C. Freeman and L. Soete 1984. Long waves, inventions and innovations. In *Long Waves in the World Economy*, C. Freeman (ed.), 63–77. Guildford: Butterworth.

Clarke, J. 1981. Harwell's heritage. *Harlequin* **24**(1), 5–8.

Cockcroft, J. (ed.) 1965. *The organization of research establishments*. Cambridge: Cambridge University Press.

Cole, H. N. 1980. *The story of Aldershot: a history of the civil and military towns*. Aldershot: Southern Books (Aldershot).

Cullingworth, J. B. 1979. *Environmental planning 1939–1969*. Vol. 3, *New towns policy*. London: HMSO.

Debenham, Tewson and Chinnocks 1983. *High tech: myths and realities*. London: Debenham, Tewson and Chinnocks.

Drake, J., H. L. Yeadon and D. I. Evans 1969. *Motorways*. London: Faber & Faber.

Drucker, P. F. 1969. *The age of discontinuity: guidelines to our changing society*. London: Heinemann.

Duchin, F. 1983. Economic consequences of military spending. *Journal of Economic Issues* **17**, 543–53.

Duckworth, G. H. 1895. Scientific, surgical and electrical instruments. In *Life and labour of the people in London*, C. Booth (ed.), Vol. 6, 35–53. London: Macmillan.

Dunne, J. P. and R. P. Smith 1984. The economic consequences of reduced UK military expenditure. *Cambridge Journal of Economics* **8**, 297–310.

Economist 1982. Britain's Sunrise Strip. *The Economist*, 30 January.

EITB (Electrical Industry Training Board) 1984. *Does the M4 Corridor exist?* Research Note. London: EITB.

Electrical and Radio Trading 1947. *'Electrical and Radio Trading' buyers' review and directory 1947*. London: Electrical and Radio Trading.

Ellin, D. and A. Gillespie 1983. *High technology product industries and job creation in Great Britain*. Working Note, CURDS. University of Newcastle: Centre for Urban and Regional Development Studies.

Feder, B. J. 1983. Britain's science corridor. *New York Times*, 24 April.

Feldman, M. 1985. Biotechnology and local economic growth. In Hall and Markusen 1985, 65–79.

Firn, J. R. and D. Roberts 1984. High technology industries. In *Industry, policy and the Scottish economy*, N. Hood and S. Young (eds), 288–325. Edinburgh: Edinburgh University Press.

Fogarty, M. P. 1945. *Prospects of the industrial areas of Great Britain*. London: Methuen.

Foley, D. 1963. *Controlling London's growth: planning the Great Wen 1940–1960*. Berkeley and Los Angeles: University of California Press.

Forrester J. 1985. *Defence Sector Procurement Opportunities for High Technology Small Firms*. London: Small Business Research Trust.

Forshaw, J. H. and P. Abercrombie 1943. *County of London plan 1943*. London: Macmillan.

Fothergill, S. and G. Gudgin 1982. *Unequal growth: urban and regional employment change in the UK*. London: Heinemann Educational.

Freeman, C. 1984. *The role of technical change in national economic development*. Paper presented at the Workshop on Regional Perspectives on Technological Change, Small Firms and Employment. University of Newcastle upon Tyne, Centre for Urban and Regional Development Studies.

Freeman, C., J. Clark and L. Soete 1982. *Unemployment and technological innovation: a study of long waves and economic development*. London: Frances Pinter.

Freiberger, P. and M. Swaine 1984. *Fire in the valley: the making of the personal computer*. Berkeley, Calif.: Osborne/McGraw-Hill.

Fuller Peiser 1985. *High technology '85*. London: Fuller Peiser.

GB ACARD (Advisory Council for Applied Research and Development) 1979. *Technological change*. London: HMSO.

GB AERE (Atomic Energy Research Establishment) 1980. *Harwell: research laboratory of the United Kingdom Atomic Energy Authority*. Harwell: AERE.

GB Air Ministry 1943. *British Overseas Airways Corporation. Further correspondence: February–March 1943* (Cmd. 6442), (BPP 1942–43, 7). London: HMSO.

GB Board of Trade 1934–9. *Survey of industrial development, 1933 (etc.): particulars of factories opened, extended and closed in 1933 (etc.) with some figures for 1932 (etc.)* London: HMSO.

GB Cabinet Office 1984. *Annual review of government funded R&D*. London: HMSO.

GB Defence Committee 1981. *Ministry of Defence organisation and procurement* (BPP 1981–2, 22). London: HMSO.

GB Department of the Environment 1980. *Development control – policy and practice*. Circular 22/80. London: DOE.

GB Department of the Environment 1983. *Industrial development: consultative document*. London: DOE.

GB Department of the Environment 1984. *Industrial development*. Circular 16/84. London: DOE.

GB DSIR (Department of Scientific and Industrial Research) 1930. *Report of the Radio Research Board for the period ended 31 March 1929*. London: HMSO.

GB DSIR (Department of Scientific and Industrial Research) 1951. *NPL. Jubilee book of the National Physical Laboratory*, by John Langdon-Davies. London: HMSO.

GB Electronics EDC 1973. *Annual statistical survey of the electronics industry, 1972*. London: NEDO.

GB Estimates Committee 1965. *Seventh report, Session 1964–65: electrical and electronic equipment for the services*. HC Paper 358 (BPP 1964–65, 7). London: HMSO.

GB House of Lords (Select Committee on Science and Technology) 1983. *Engineering research and development*. HC Papers 89. Vol 1. London: HMSO.

GB Minister of Civil Aviation 1953. *London's airports* (Cmd. 8902), (BPP 1952–53, 24). London: HMSO.

GB Ministry of Civil Aviation 1945. *British Air Services* (Cmd. 6712), (BPP 1945–46, 15). London: HMSO.

GB Ministry of Civil Aviation 1946. *London Airport: report of layout panel*. London: HMSO.

GB Ministry of Civil Aviation 1948. *London Airport*. London: HMSO.

GB Ministry of Defence 1984a. *Proposed development at the Clyde submarine base, environmental impact assessment*. London: MOD and Property Services Agency.

GB Ministry of Defence 1984b. *Statement on the defence estimates 1984*. Cmnd. 9227. London: HMSO.

GB Ministry of Defence 1985. Statement on the defence estimates. Cmnd 9430. London: HMSO.

GB Ministry of Defence n.d. *Selling to the MOD*. London: Ministry of Defence.

GB Ministry of Housing and Local Government 1964. *The South East study 1861–1981*. London: HMSO.

GB Ministry of Supply 1952. *Harwell: the British Atomic Research Establishment 1946–1951*. London: HMSO.

GB Ministry of Supply 1954. *Britain's atomic factories: the story of atomic energy production in Britain*, by K.E.B. Jay. London: HMSO.

GB National Economic Development Council 1983. *Civil exploitation of defence technology*, by I. Maddock. London: NEDC.

GB Parliament 1951. *Defence programme. Statement made by the Prime Minister in the House of Commons on Monday, 29 January, 1951*. London: HMSO.

GB RAE (Royal Aircraft Establishment) 1955. *Fifty years at Farnborough*. London: HMSO.

GB RC (Royal Commission) Distribution of the Industrial Population 1938. *Minutes of evidence, tenth and eleventh days, Wednesay, 2nd February and Thursday, 3rd February, 1938*. London: HMSO.

GB RC (Royal Commission) Distribution of the Industrial Population 1940. *Report*. Cmd. 6153. London: HMSO.

GB SC (Select Committee on) Estimates 1947. *Third report from the Select Committee on Estimates, session 1946–47: expenditure on research and development*. HC Paper 132–1 (BPP 1946–47, 6). London: HMSO.

GB SC (Select Committee on) Estimates 1951. *Third report from the Select Committee on Estimates, session 1950–51: rearmament*. HC Paper 178 (BPP 1950–51, 5). London: HMSO.

GB SC (Select Committee on) Estimates 1953. *Seventh report from the Select Committee on Estimates, session 1952–53: rearmament*. HC Paper 178 (BPP 1952–53, 4). London: HMSO.

GB SC (Select Committee on) Estimates 1956. *Seventh report from the Select Committee on Estimates, session 1955–56: naval research and development*. HC Paper 345 (BPP 1955–56, 8). London: HMSO.

GB SC (Select Committee on) Science (and Technology) 1969. *Second report: defence research*. HC Papers 213 (BPP 1968–69, 21). London: HMSO.

GB South East Economic Planning Council 1967. *A strategy for the South East*. London: HMSO.

GB South East Joint Planning Team 1970. *Strategic plan for the South East: report by the South East Joint Planning Team*. London: HMSO.

GB South East Joint Planning Team 1976. *Strategy for the South East: 1976 review: report with recommendations by the South East Joint Planning Team*. London: HMSO.

GB Steering Group on Research and Development Establishments 1980. *Report* (Strathcona report). London: MOD.

GB Treasury 1984. *The next ten years: public expenditure and taxation into the 1990s*. Cmnd 9189. London: HMSO.

GB UKAEA (United Kingdom Atomic Energy Authority) 1958. *A brief guide: Harwell: the Atomic Energy Research Establishment*. London: UKAEA.

GB UKAEA (United Kingdom Atomic Energy Authority) 1959. *Atomic Weapons Research Establishment at Aldermaston*. London: UKAEA.

Gershuny, J. and I. Miles 1983. *The new service economy: the transformation of employment in industrial societies*. London: Frances Pinter.

Gibbs, D. 1983. *The spatial incidence of high technology industry and technological change*. Working Note. Newcastle upon Tyne: University, Centre for Urban and Regional Development Studies.

Glasmeier, A. 1985. Innovative manufacturing industries: spatial incidence in the United States. In Castells 1985, 55–80.

Glasmeier, A., P. Hall and A. Markusen 1983. *Defining high technology industries*. Working Paper 407. University of California, Berkeley: Institute of Urban and Regional Development.

Goddard, J. B. 1973. Office linkages and location. *Progress in Planning* 1, part 2.

Goddard, J. B. *et al*. 1983. Technical innovation in a regional context: empirical evidence and policy options. Paper for the workshop on 'Research, Technology and Regional Policy'. Paris: OECD.

Goddard, J. B., A. E. Gillespie, J. F. Robinson and A. T. Thwaites 1985. New information technology and urban and regional development. In *Technological change and regional development*, A. T. Thwaites and R. P. Oakey (eds). London: Frances Pinter.

Gowing, M. M. 1964. *Britain and atomic energy 1939–1945*. London: Macmillan.

Greenwood, D. 1982. The defense policy of the United Kingdom. In *The Defense Policies of Nations*, D. J. Murray and P. R. Viotti (eds), 197–225. Baltimore: Johns Hopkins University Press.

Greenwood, D. and J. Short 1973. *Military installations and local economies – a case study: the Moray air stations*. Aberdeen Studies in Defence Economics, no. 4. Aberdeen: University of Aberdeen.

Gregory, R. 1967. The minister's line: or, the M4 comes to Berkshire. *Public Administration* **45**, 113–28 and 269–86.

Grinling, C. H. 1898. *The history of the Great Northern Railway, 1845–1895*. London: Methuen.

Gummett, P. 1984. Defence research policy. In *UK research policy: critical review of policies for publicly-funded science and technology*, M. Goldsmith (ed.), 57–81. London: Longman.

Hague, D. C and J. H. Dunning 1955. Costs in alternative locations: the radio industry. *Review of Economic Studies* **22**, 203–13.

Hall, P. 1962. *The industries of London since 1861*. London: Hutchinson University Library.

Hall, P. (ed.) 1981. *The inner city in context*. London: Heinemann.

Hall, P. 1985. The geography of the fifth Kondratieff. In Hall and Markusen 1985, 1–19.

Hall, P. and A. Markusen 1983. *Technology innovation and regional economic development. A research proposal*. University of California, Berkeley. Unpublished.

Hall, P. and A. Markusen (eds) 1985. *Silicon landscapes*. London: Allen & Unwin.

Hall, P., A. Markusen, R. Osborn and B. Wachsman 1983. *The California software industry: problems and policy issues*. Working Paper no. 410. Berkeley: University of California, Institute of Urban and Regional Development.

Hall, P., A. R. Markusen, R. Osborn and B. Wachsman 1985. The California software industry: economic development prospects. In Hall and Markusen 1985, 49–64.

Hall, P., R. Thomas, H. Gracey and R. Drewett 1973. *The containment of urban England*. 2 volumes. London: Allen & Unwin.

Hampshire, County of 1978. *North East Hampshire structure plan: submitted 1978*. Winchester: Hampshire County Council.

Hampshire County Council 1984. *High technology industry in Hampshire*. Strategic Planning Paper, 15. Winchester: Hampshire County Council.

Harris, F. and R. McArthur 1985. The issue of high technology: an alternative view. Working Paper 16. Manchester: North West Industry Research Unit, School of Geography, University of Manchester.

Hart, D. A. 1976. *Strategic planning in London: the rise and fall of the primary road network*. Oxford: Pergamon.

Helfgott, R. B. 1959. Women's and children's apparel. In *Made in New York*, M. Hall (ed.), 19–34. Cambridge, Mass.: Harvard University Press.

Henderson, J. and A. J. Scott 1987. The growth and internationalisation of the American semiconductor industry: labour process and the changing spatial organisation of production. In *High Technology Industry: An International Survey*, M. Breheny and R. McQuaid (eds). Beckenham: Croom Helm.

Hirsch, S. 1967. *Location of industry and international competitiveness*. Oxford: Clarendon Press.

Hoover, E. M. 1948. *The location of economic activity*. New York: McGraw-Hill.

Hornby, W. 1958. *Factories and plant*. History of the Second World War, United Kingdom Civil Series. London: HMSO and Longmans, Green.

Howells, J. R. L. 1984. The location of research and development: some observations and evidence from Britain. *Regional Studies* **18**, 13–30.

Hymer, S. H. 1975. The multinational corporation and the law of uneven development. In *International firms and modern imperialism*, H. Radice (ed.), 37–62. Harmondsworth: Penguin.

IPCS (Institute of Professional Civil Servants) 1984. *Nuclear arms, defence spending and jobs*. Mimeo. London: ICPS.

James, A. G. T. 1976. *The Royal Air Force: the past 30 years*. London: Macdonald & Jane's.

Jones, R. V. 1978. *Most secret war*. London: Hamish Hamilton.

Jones, S. R. 1982. *An accessibility analysis of the impact of the M25 motorway*. TRRL Laboratory Report 1055. Crowthorne: Transport and Road Research Laboratory.

Kaldor, M. 1982. *The baroque arsenal*. London: Deutsch.

Keeble, D. E. 1968. Industrial decentralization and the metropolis: the north-west London case. *Transactions and Papers, Institute of British Geographers* **44**, 1–54.

Kondratieff, N. D. 1935. The long waves in economic life. *Review of Economic Statistics* **17**, 105–15.

Kuznets, S. S. 1940. Schumpeter's business cycles. *American Economic Review* **30**, 250–71.

Kuznets, S. S. 1946. *National product since 1869*. NBER Publications, no. 46. New York: National Bureau of Economic Research.

Kuznets, S. S. 1966. *Modern economic growth: rate, structure and spread*. New Haven and London: Yale University Press.

Langridge, R. J. 1984. *Defining 'high-tech' for locational analysis*. Discussion Paper 22, Series C. Reading: University of Reading, Department of Economics.

Law, C. M. 1983. The defence sector in British regional development. *Geoforum* **14**, 169–84.

Lichfield, N. and Partners, Goldstein Leigh Associates 1981. *M25 London orbital: property market effects*. London: Nathaniel Lichfield and Partners and Goldstein Leigh Associates.

Lighthill, M. J. 1965. The Royal Aircraft Establishment. In Cockcroft 1965, 28–54.

Little, A. D. 1977. *New technology-based firms in the UK and the Federal Republic of Germany*. A Report for the Anglo-German Foundation for the Study of Industrial Society. London: Anglo-German Foundation.

Llewellyn Smith, H. (ed.) 1931. *New survey of London life and labour*, Vol. 2. London: P. S. King.

Lösch, A. 1954 (1944). *The economics of location*. Translated by W. H. Woglom. New Haven: Yale University Press.

Lovering, J. 1983. *Regional intervention. Defence industries and the structuring of space in Britain: the case of Bristol and South Wales*. Paper given to the Cardiff Branch, Regional Studies Association.

Lovering, J. 1984. *The development of the aerospace industry in Bristol 1910–1984*. Project Working Paper 5, ESRC Inner City in Context Research Project. University of Bristol: School for Advanced Urban Studies.

MacDermot, E. T. 1927. *History of the Great Western Railway*, 2 vols in 3 parts. London: Great Western Railway Company.

McKie, D. 1973. *A sadly mismanaged affair: a political history of the third London airport*. London: Croom Helm.

Malecki, E. J. 1980a. Corporate organization of R and D and the location of technological activities. *Regional Studies* **14**, 219–34.

Malecki, E. J. 1980b. Science and technology in the American urban system. In *The American Metropolitan System: Past and Future*, S. D. Brunn and J. O. Wheeler (eds), 127–44. London: Edward Arnold.

Malecki, E. J. 1980c. Technological change: British and American research themes. *Area* **12**, 253–60.

Malecki, E. J. 1981a. Public and private sector interrelationships, technological change, and regional development. *Papers of the Regional Science Association* **47**, 121–38.

Malecki, E. J. 1981b. Government-funded R&D: some regional economic implications. *Professional Geographer* **33**, 72–82.

Mandel, E. 1975. *Late capitalism*. London: Verso.

Mandel, E. 1980. *Long waves of capitalist development*. Cambridge: Cambridge University Press.

Markusen, A. 1984. *High technology and the future of employment*. Paper presented at the National Conference on Economic Dislocation and Job Loss. Washington, DC, 9–10 April 1984.

Markusen, A. 1985. *Profit cycles, oligopoly and regional development*. Cambridge, Mass.: MIT Press.

Markusen, A., P. Hall and A. Glasmeier 1986. *High-tech America: the what, how, where and why of the sunrise industries*. Boston: Allen & Unwin.

Marshall, A. 1920 (1890). Principles of economics. London: Macmillan.

Martin, J. E. 1966. *Greater London: an industrial geography*. London: Bell.

Massey, D. 1984. *Spatial divisions of labour: social structures and the geography of production*. London: Macmillan.

Massey, D. 1985. Which 'new technology'? In Castells 1985, 302–16.

Massey, D. and R. Meegan 1982. *The anatomy of job loss: the how, why and where of employment decline*. London: Methuen.

Mensch, G. 1979. *Stalemate in technology: innovations overcome the depression*. Cambridge, Mass.: Ballinger.

Miller, H. 1963. *The way of enterprise: a study of the origins, problems and achievements of post-war British firms*. London: André Deutsch.

National Economic Development Council 1983. *The introduction of new technology*. London: NEDC.

Neil, A. 1983. The information revolution. *The Listener*, 23 June.

Nock, O. S. 1980. *Two miles a minute: the story behind the conception and operation of Britain's High Speed and Advanced Passenger Trains*. Cambridge: Patrick Stephens.

Northcott, J. and P. Rogers 1984. *Microelectronics in British industry: the patterns of change*. London: Policy Studies Institute.

Oakey, R. P. 1981. *High-technology industrial location: the instruments example*. Aldershot: Gower.

Oakey, R. P. 1983. *Research and development cycles, investment cycles and regional growth*. Newcastle upon Tyne, Centre for Urban and Regional Development Studies, Discussion Paper no. 48.

Oakey, R. P., A. T. Thwaites and P. A. Nash 1980. The regional distribution of innovative manufacturing establishments in Great Britain. *Regional Studies* **14**, 235–53.

Pearson, F. S. 1983. The question of control in British defense sales policy. *International Affairs* **59**, 211–38.

Perez, C. 1983. Structural change and assimilation of new technologies in the economic and social systems. *Futures* **15**, 359–75.

Perry, D. C. and A. J. Watkins 1978. *The rise of the sunbelt cities*. Urban Affairs Annual Reviews, 14. Beverly Hills and London: Sage.

Phillips, B. *et al.* 1983. Western corridor: a special report. *The Times*, 30 June.

Pite, C. 1980. Employment and defence. *Statistical News* **51**, 15–20.

Postan, M. M. 1952. *British war production*. History of the Second World War, United Kingdom Civil Series. London: HMSO and Longmans, Green.

Postan, M. M., D. Hay and J. D. Scott 1964. *Design and development of weapons: studies in governmental and industrial organisation*. History of the Second World War, United Kingdom Civil Series. London: HMSO and Longmans, Green.

Powell, A. G. 1960. The recent development of Greater London. *Advancement of Science* **17**, 76–86.

Premus, R. 1982. *Location of high-technology firms and regional economic development*. US Congress, Joint Economic Subcommittee on Monetary and Fiscal Policy. Washington, DC: GPO.

Pyatt, E. 1983. *The National Physical Laboratory: a history*. Bristol: Adam Hilger.

Reading Borough Council 1983. *Central Reading district plan: written statement*. Reading: Reading Borough Council.

Rees, J. 1979. Technological change and regional shifts in American manufacturing. *Professional Geographer* **31**, 45–54.

Robertson, J. A. S., J. M. Briggs and A. Goodchild 1982. *Structure and employment prospects of the service industries*. Department of Employment Research Paper no. 3. London: Department of Employment.

Rogers, E. M. and J. K. Larson 1984. *Silicon Valley fever: growth of high-technology culture*. New York: Basic Books.

Rolt, L. T. C. 1970. *Isambard Kingdom Brunel*. Harmondsworth: Penguin.

Rothwell, R. 1982. The role of technology in industrial change: implications for regional policy. *Regional Studies* **16**, 361–9.

Rothwell, R. and W. Zegveld 1981. *Industrial innovation and public policy: preparing for the 1980s and 1990s*. London: Frances Pinter.

Saunders, H. St G. 1954. *Royal Air Force, 1939–45*. Vol. 3, *The fight is won, 1943–45*. London: HMSO.

Saxenian, A. 1981. *Silicon chips and spatial structure*. Berkeley: University of California, Institute of Urban and Regional Development, Working Paper No. 345.

Saxenian, A. 1985a. The genesis of Silicon Valley. In Hall and Markusen 1985, 20–34.

Saxenian, A. 1985b, Silicon Valley and Route 128: regional prototypes or historic exceptions? In Castells 1985, 81–105.

Sayer, A. and K. Morgan 1987. High technology industry and the international division of labour: the case of electronics. In *High technology industry: an international survey*, M. Breheny and R. McQuaid (eds). Beckenham: Croom Helm.

Sayer, R. A. 1985. Industry and space: a sympathetic critique of radical research. *Environment and Planning D: Society and Space* **3**, 3–29.

Schumpeter, J. A. 1961 (1911). *The theory of economic development*. Cambridge, Mass: Harvard University Press.

Schumpeter, J. A. 1982 (1939). *Business cycles: a theoretical, historical and statistical account of the capitalist process*. 2 volumes. New York and London: McGraw-Hill. Reprinted Philadelphia: Porcupine Press.

SCLSERP (Standing Conference on London and South East Regional Planning) 1982. *The impact of the M25*. SC 1706. London: SCLSERP.

SCLSERP (Standing Conference on London and South East Regional Planning) 1983. *Employment Monitor for 1982/3*, SC1812. London: SCLSERP.

Segal Quince and Partners 1985. *The Cambridge phenomenon: the growth of high technology industry in a university town*. Cambridge: Segal Quince and Partners.

SERPLAN (Standing Conference on London and South East Regional Planning) 1985. *Regional trends in the South East: the South East regional monitor 1984–85*. RPC 369. London: SERPLAN.

Short, J. 1981a. Defence spending in the UK regions. *Regional Studies* **15**, 101–10.

Short, J. 1981b. *Public expenditure and taxation in the UK regions*. London: Gower.

Short, J., T. Stone and D. Greenwood 1974. *Military installations and local economies – a case study: the Clyde submarine base*. Aberdeen Studies in Defence Economics, no. 5. Aberdeen: University of Aberdeen.

Smith, D. H. 1933. *The industries of Greater London: being a survey of the recent industrialisation of the northern and western sectors of Greater London*. London: P. S. King.

Smith, F. W. F. 1961. *The Prof in two worlds: the official life of Professor F. A. Lindemann, Viscount Cherwell*. London: Collins.

Starkie, D. 1982. *The motorway age: road and traffic policies in post-war Britain*. Oxford: Pergamon.

Stoddart, R. 1983. Structure plans in Berkshire – theory and practice. In *English Structure Planning*, D. T. Cross and M. R. Bristow (eds). London: Pion.

Street, S. and J. Beasley 1985. *Investing in Defence*. London: de Zoete Bevan.

Study Team 1975. *Reading, Wokingham, Aldershot, Basingstoke subregional study: report of the Study Team*. Reading: Berkshire County Council.

Sutherland, G. 1965. The National Physical Laboratory. In Cockcroft 1965, 6–27.

Thomas, D. 1983a. How would defence cuts hit jobs? *New Society* **63**, 334–6.

Thomas, D. 1983b. England's golden west. *New Society* **64**, 177–8.

Thwaites, A. T. 1982. Some evidence of regional variations in the introduction and diffusion of industrial processes within British manufacturing industry. *Regional Studies* **16**, 371–81.

Todd, D. 1980. The defence sector in regional development. *Area* **12**, 115–21.

Todd, D. 1981. Regional variations in naval construction. The British experience, 1895–1966. *Regional Studies* **15**, 123–42.

Törnqvist, G. 1968. Flows of information and the location of economic activity. *Geografiska Annaler* **508**, 99–107.

Udis, B. 1978. *From guns to butter: technology organisations and reduced military spending in Western Europe*. Cambridge, Mass.: Ballinger.

United States Department of Commerce 1983. *An assessment of the U.S. competitiveness in high technology industries*. Washington, DC: International Trade Administration, US Department of Commerce.

van Duijn, J. J. 1983. *The long wave in economic life*. London: Allen & Unwin.

Vernon, R. 1966. International investment and international trade in the product cycle. *Quarterly Journal of Economics* **80**, 190–207.

Vick, F. A. 1965. The Atomic Energy Research Establishment, Harwell. In Cockcroft 1965, 55–77.

von Thünen, J. H. 1966 (1826). *Von Thünen's isolated state*. Edited by P. Hall, translated by C. M. Wartenberg. Oxford: Pergamon.

Walker, P. B. 1971, *Early aviation at Farnborough: the history of the Royal Aircraft Establishment*. Vol. 1: *Balloons, kites and airships*. London: Macdonald.

Walker, P. B. 1974. *Early aviation at Farnborough: the history of the Royal Aircraft Establishment*. Vol. 2: *The first aeroplanes*. London: Macdonald.

Warren, J. P. 1982. *The 50-year boom–bust cycle: the case for Kondratieff's long wave theory*. Godalming, Surrey: Warren, Cameron.

Watson-Watt, R. 1957. *Three steps to victory: a personal account by radar's greatest pioneer*. London: Odhams.

Weber, A. 1929 (1909). *Theory of the location of industry*. Translated by C. J. Friedrich. Chicago: University of Chicago Press.

Whitehouse, P. and D. S. Thomas 1984. *The Great Western Railway: 150 glorious years*. Newton Abbot: David & Charles.

Whitlock, R. 1976. *Wiltshire*. London: Batsford.

Wise, M. J. 1949. On the evolution of the jewellery and gun quarters in Birmingham. *Institute of British Geographers, Transactions and Papers* **15**, 57–72.

Wright, A. J. 1983. *British airports*, 2nd edn. London: Ian Allan.

Index